D1010573

THE STEPHEN BECHTEL FUND

IMPRINT IN ECOLOGY AND THE ENVIRONMENT

The Stephen Bechtel Fund has

established this imprint to promote

understanding and conservation of

our natural environment.

Tracks and Shadows

*The publisher gratefully acknowledges the generous
contribution to this book provided by
the Stephen Bechtel Fund.*

Tracks and Shadows

FIELD BIOLOGY AS ART

Harry W. Greene

UNIVERSITY OF CALIFORNIA PRESS

BERKELEY LOS ANGELES LONDON

University of California Press, one of the most distinguished university presses in the United States, enriches lives around the world by advancing scholarship in the humanities, social sciences, and natural sciences. Its activities are supported by the UC Press Foundation and by philanthropic contributions from individuals and institutions. For more information, visit www.ucpress.edu.

University of California Press
Berkeley and Los Angeles, California

University of California Press, Ltd.
London, England

© 2013 by The Regents of the University of California

Library of Congress Cataloging-in-Publication Data

Greene, Harry W., 1945–.
    Tracks and shadows : field biology as art / Harry W. Greene.
        pages   cm.
    Includes bibliographical references and index.
    ISBN 978-0-520-23275-4 (hardback) — ISBN 978-0-520-95673-5 (ebook)
    1. Greene, Harry W., 1945–   2. Biologists—United States—Biography.
3. Nature.
    QH31.G73   2013
    570.92—dc23
    [B]   2013020395

Manufactured in the United States of America

22   21   20   19   18   17   16   15   14   13
10   9   8   7   6   5   4   3   2   1

In keeping with a commitment to support environmentally responsible and sustainable printing practices, UC Press has printed this book on Natures Natural, a fiber that contains 30% post-consumer waste and meets the minimum requirements of ANSI/NISO Z39.48-1992 (R 1997) (*Permanence of Paper*).

For Kelly, my beautiful and fearless companion . . .

CONTENTS

PART THREE
PRETTY IN SUNLIGHT

ILLUSTRATIONS

ACKNOWLEDGMENTS

Natural history is woefully underfunded, despite timeless value, so I begin by thanking five farsighted individuals, with hopes others will follow their lead. Annie Alexander, who founded the Museum of Vertebrate Zoology, was a sugar fortune heiress, explorer, big-game hunter, specimen collector, and visionary. From 1978 to 1998 her endowment provided core support for my research and teaching. Between 1997 and 2009, thanks to Canadian philanthropists Allan Stone and Susan Gibson's Lichen Foundation—without proposals or reports!—I visited far-flung locales, replaced worn-out equipment, advanced a controversial conservation initiative, felt rewarded for a heavy teaching load, and sponsored a Mexican student studying his native fauna. Allan and Susan became dear friends too, offering the sort of nonjudgmental advice one cherishes all the more as years go by. Last but not least, Dick and Joanne Bartlett's Thinking Like a Mountain Foundation provided a beautiful Chihuahuan Desert setting for writing, and our conversations as well as their books (Bartlett and Krieger 1988; Bartlett and Williamson 1995) sustained my enthusiasm for Aldo Leopold's vision of a wild future.

Friends old and new provided information on everything from youthful escapades to snake biology, along with reviewing chapters, facilitating my travels, and broadening my worldview. With apologies to anyone forgotten and for not specifying individual contributions, I thank Ed Acuña, Kraig Adler, Sofia Agudelo, Graham Alexander, Richard Allen, Ronn Altig, Melissa Amarello, John Anderson, Tom Anton, Brian Aquadro, Sir David Attenborough, Josiah Austin, Valer Austin, Randy Babb, Bruce Babbitt, Mary Bailey, Warner Bailey, Gary Baldwin, Lorraine Baldwin, Liz Balko, Dave Barker, Tony Barnosky, Steve Barten, Puja Batra, Aaron Bauer, Joel Berger, Bob Bezy, Chigiy Binell, Bernd Blossey, Christine Bogdanowitz,

Steve Bogdanowitz, Sarah Boone, Dee Bowen, Carolyn Boyd, Beth Braker, Cinthia Brasileiro, Chris Brochu, Butch Brodie III, Dan Brooks, Jim Brown, Gordon Burghardt, Young Cage, Jonathan Campbell, John Carothers, Chuck Carpenter, Geoff Carpenter, David Cerasale, Dee Ann Chamberlain, Lauren Chan, Mark Chow, Jon Christensen, David Clark, Deborah Clark, Don Clark, Susan Clayton, Katie Colbert, Jay Cole, Tim Cole, Rob Colwell, Richard Conniff, Bob Corley, Margie Crisp, David Cundall, Chris Cuomo, Ralph Cutter, Shannon Davies, Mac Davis, Gage Dayton, Paul Dayton, Dan Decker, Arya Degenhardt, Alex Deufel, Tom Devitt, Tish Diffenbaugh, Josh Donlan, Mo Donnelly, Bob Drewes, Hugh Drummond, Bill Duellman, Beverly Dugan, Rob Dunn, Sandy Echternacht, Cody Edwards, Tom Eisner, Steve Emlen, Jim Estes, Chris Feldman, Lee Fitzgerald, Courtney Fitzpatrick, John Fitzpatrick, Sarah Fitzpatrick, Tom Fleischner, Michael Fogden, Patricia Fogden, Susan Fogden, Mercedes Foster, Laurence Frank, Carl Franklin, Caroline Fraser, Joe Furman, Marilia Gaiarsa, Gabe Gartner, Tom Gavin, Monica Geber, Cameron Ghalambor, Graeme Gibson, Susan Gibson, Larry Gilbert, Chris Giorni, Warner Glenn, Wendy Glenn, Matt Goode, George Gorman, Carol Gould, Marco Granzinolli, Nancy Greig, Mike Grundler, Celio Haddad, Kim Haines-Eitzen, Laurie Hall, Winnie Hallwachs, Mary Ann Handel, Jill Hanna, Bob Hansen, Jonathan Hanson, Roseann Hanson, Billie Hardy, David Hardy Sr., Matt Hare, Jim (snake venom) Harrison, Rick Harrison, Larry Hatteberg, Alden Hayes, Karen Hayes, Tom Headland, Tracy Heath, Eileen Hebets, Linda Hedges, Bob Henderson, Christie Henry, Hal Herzog, Troy Hibbits, Mary Lou Higgins, Ann Hillis, David Hillis, Justin Hite, Iris Holmes, Andy Holycross, Ron Hoy, Ray Huey, Mary Hufty, Lynne Isbell, Kate Jackson, Elliott Jacobson, Fabián Jaksic, Dan Janzen, Chris Jenkins, Jay Johnson, Bob Jones, Tom Jones, Jan Kalina, Yula Kapetanakos, Peter Kareiva, John Karges, Nancy Karraker, Nate Kley, Mimi Koehl, Walt Koenig, Alex Krohn, Travis LaDuc, Bill Lamar, Manuel Leal, Jeff Lemm, Estella Leopold, Harvey Lillywhite, Irby Lovette, Claudia Luke, Holly Lutz, Don Lyman, Sherry Machado, Steve Mackessy, Katie Mackinnon, Midge Marchaterra, Bryan Maritz, Otavio Marques, Emma Marris, Laura Martin, Paul Martin, Tom Martin, Marcio Martins, Ken McCloud, Robert McCurdy, Roy McDiarmid, Bill McGrew, Mike McRae, Katie Meier, Gerold Merker, Ardell Mitchell, Mark Moffett, Bill Montgomery, Patricia Morellato, Jim Morin, Josh Morse, Jen Moslemi, Caterina Myers, Shahid Naeem, Cristiano Nogueira, Erika Nowak, Kira Od, Pete Oxford, Craig Packer, Barbara Page, Chris Parkinson,

Gabriela Parra, Hank Paulson, Wendy Paulson, Greg Pauly, John Pearse, Gina Pecararo, Bobbi Peckarsky, Tony Phelps, Bryan Piazza, Louis Porras, Cynthia Prado, Jill Pruetz, Bill Pyburn, Karen Pyburn, Wanda Pyburn, George Rabb, Sarah Rabkin, Rob Raguso, Phil Ralidis, Bob Reed, Stephanie Remmington, Van Remsen, Art Rentfro, Roger Repp, Alan Resetar, Jon Richmond, Ruth Rigler, Jesús Rivas, Jeanne Robertson, Carmen Rojas, Dick Root, Jessica Rothman, Dan Rubenstein, Dustin Rubenstein, Becca Safran, Manuel Santana, Anna Savage, Alan Savitzky, Ricardo Sawaya, Ivan Sazima, Michael Schaer, George Schaller, Chris Schneider, Gordon Schuett, Alexis Schuler, Kurt Schwenk, Jim Shaw, Wade Sherbrooke, Rick Shine, Jake Shrum, Elizabeth Shull, Jesús Sigala, Jack Sites, Dave Skelly, Art Smith, Polly Smith-Blackwell, Gary Snyder, Alejandro Solórzano, Jed Sparks, Josh Stafford, Barbara Stein, Sam Stephenson, Ann Stingle, Will Stolzenburg, Allan Stone, Jay Storz, Mary Sunderland, Mel Sunquist, Hilary Swain, Jim Tantillo, Emily Taylor, Lonn Taylor, Nat Taylor, Stan Temple, Tereza Thomé, Steve Tilley, Kristine Tollestrup, Barney Tomberlin, Joe Truett, David Tutt, Guillermo Velo-Antón, Jens Vindum, Ted Vitali, Rob Voss, Corine Vriesendorp, Marvalee Wake, David Wake, Mary Jane West-Eberhard, Josh Whorley, Kristen Wiley, Johnny Williams, Pam Williams, Kevin Wiseman, Wolfgang Wüster, Janet Wylde, Anne Yoder, Bruce Young, Fernando Zara, Bob Zink, and George Zug.

A few folks deserve special thanks. Will, his wife, Candace, and our mom were especially helpful in reconstructing—often deconstructing!—my youth. Virginia and Henry Fitch cooperated enthusiastically, as did their daughter and son-in-law, Alice and Tony Echelle. The late Jim Clark signed me on for a second book with the University of California Press. Writers Bob Benson, Chuck Bowden, Jennie Dusheck, Jim Harrison, BK Loren, and Jeri Spann gave critical feedback and encouragement. Marty Crump, Darragh Hare, Mike Lannoo, Jonathan Losos, Hugh McCrystal, Joe Mendelson, and Kelly Zamudio commented on the entire manuscript, as did editors Anne Canright, Blake Edgar, Doris Kretschmer, and Rose Vekony for the press.

Grateful acknowledgment is made to the following organizations and individuals for permission to include revised excerpts of my own previously published work (see the Bibliography for full citations): Greene 1997: The Regents of the University of California; 1999b, 2009: The Society for the Study of Amphibians and Reptiles; 2003: Dave Foreman, publisher of *Wild Earth;* 2010: The American Association for the Advancement of Science; and 2011: Roberts and Company Publishers.

PART ONE

Descent with Modification

ONE

Tracks and Shadows

A TIMELY OLD QUOTE SIDLES around the two entwined themes of this book, my eccentric meditation on natural history. Writing from 1849 about Sierra Nevada streams devastated by gold miners, journalist Bayard Taylor likened nature to "a princess, fallen into the hands of robbers, who cut off her fingers for the sake of the jewels she wears."[1] His brutal imagery frames a modern dilemma, because although many people believe animals relocate when their habitats are destroyed, most organisms have nowhere to go. They will die rather than move. Worse yet, these losses are usually unseen and writ large all over the world, so we truly are thieves, pillaging the future. My first theme, coming to grips with this predicament, challenges everyone who cares about biodiversity—even if the effort to clarify what we want turns out to be a philosophical snake in the grass, more nuanced and elusive than I long supposed.

Taylor's slashing tone also resonates with a second theme, the twists and turns of my personal quest for wildness. Early on, as a curious youngster with rural grandparents, I discovered the seductive joys of nature study. From horned lizards and livestock on a Texas farm to elephants and lions in zoos, I watched and wondered. How can they eat only ants or hay or meat? Why do cow patties look different from horse dung, and do insects poop? I picked up a box turtle, peered into scarlet eyes as the head craned out, and asked my mother, "Where are the ears?" In a child's naive but earnest way, I yearned to reveal their secrets, and later, as I read and traveled widely, grander questions caught my attention: Why are some animals similar and others different? Why are there so many species in the tropics? And as the human population climbs on past seven billion, will future generations still marvel at nature?

There has followed a lifetime of chasing serpents, mostly real but occasionally metaphorical. From earliest memories until age thirteen, I aspired to be a cowboy or an explorer. Then I met a zoology professor and vowed to become an academic. Soon I joined several organizations for herpetologists—folks interested in amphibians and reptiles, informally called herpers—and began publishing in their journals, obsessed with biology. In college, because my late-blooming, all-consuming social life required cash, I hired on at a local funeral parlor as a mortician's assistant and ambulance driver. Boy Scout training and vague notions of what the job would entail were my only qualifications, but I relished the excitement. It was only natural, then, that after graduation, having barely managed passing grades as a biology major, I would enlist as an army medic.

During my twenties I helped hundreds of ill and injured people, as well as watched a dozen or so die from shootings, stabbings, and accidents. I stitched up autopsied bodies, was bitten by an epileptic and squirted by severed arteries, had an assailant turn on me with a knife and listened uneasily to nearby gunfire. One night I tried to save a toddler with an allergic drug reaction, and forty-five years later, defenses softened by a good cabernet, I still suffer the agony in her mother's screams. There were happy endings, too—just days after that little girl died I placed a squirming newborn at the breast of another young woman. Through it all, field biology was a respite, and by the time thirty rolled around I was back on track for a Ph.D., studying snake evolution and utterly clueless as to how those experiences might illuminate the issues with which this book is concerned.[2]

A decade later and tenured at Berkeley, I'd lost an undergraduate advisor and a lover to murders, a heart attack had dropped my father, and several friends had died way too young. Their deaths provoked sensations of choking on explosions and desperate grappling, as if I might strangle reality back to the future, and at times these people seemed oddly still present, like phantoms of amputated limbs. Personal frailty intruded too, during those middle years, in episodes I'd gladly never repeat. Some were exhilarating yet over so fast they remained emotionally obscure, as when a high-speed collision spun our pickup truck through the rainy night into an Arizona pasture, or my foot bumped a Central American bushmaster, jolting the giant viper and me into mutually favorable defensive responses. Other threats loomed more ominous with every endless minute, like when we sat in speechless terror while our Aeroperú jet, one engine streaming flames, circled back to Lima, or were confronted by angry, armed men in Uganda.

Little wonder, given those brushes with mortality, that desert writer Ed Abbey's admonition to "throw metaphysics to the dogs, I never heard a mountain lion bawling over the fate of his soul" beckons like a Buddhist koan.[3] And perhaps it's not surprising that natural-born killers inspired me beyond scientific justification, as if confronting their deadly essence might solve more private riddles. The upshot has been rewards akin to those that motivate artists: animals are the focus of my teaching and research, but field-work has also been contemplative, inspiring me to pay attention and live more fully. The practice of natural history, I have learned, fosters peace of mind.

Predators are linked in our psyches with wildness, perhaps all the more so for those who study them. In Costa Rica fleeting shadows and strange sounds intrigued me, and because we found tracks and scats of jaguars I always hoped but never expected to see one of those great hunters. Instead I poked through droppings and identified the drab remnants of lives briefly met, puzzled over little cloven hooves of collared peccaries and scythelike claws of three-toed sloths, the scaly feet and parchment-shelled eggs of green iguanas.[4] Once a botanist led me to the bloody husk of a nine-banded armadillo, all that was left of a fresh kill. And late at night, deep in the black woods, I thought about skull-piercing canines and meat-rasping tongues, tried to imagine the prey's fine-tuned senses and gut-twitching anxieties. Do those wild-pig relatives squeal in their final moments, and would the lizards know what hit them? Could I empathize with armadillos while contending with an empty belly or hungry young-sters back in the den?

When a local entomologist grumbled, "Everybody wants to meet *el tigre*," I chimed in about ecotourists seeking a quick nature fix, as if they were rushing through the Louvre for a peek at Mona Lisa. Better be con-tent with turds and pugmarks, I mused, yearn for a glimpse of the great rainforest carnivore but settle for heightened awareness. Then, during one among countless nights searching for snakes, a companion exclaimed, "Hey, a cat!"—it had bounded across the trail in front of him—and our lights swung into the forest. The jaguar squinted from thirty feet away, all round head and broad shoulders, rosettes and long tail; just as suddenly, with not so much as a whispered paw on leaves, there were only small palms and saplings in the headlamp beams. *No more cat,* as if it had *evaporated,* and in those few seconds we more easily empathized with the Mayans, Olmecs, and others who have imbued forest creatures with mystical qualities.

Years later the memory of that animal surfaced when, with my wife, Kelly, and two Mexican friends, I backpacked from pine-oak forest that rims the four-thousand-foot-deep Barranca del Cobre (Copper Canyon) down into sweltering tropical thorn scrub along the Rio Urique. Fox scats and other carnivore signs were common along the canyon's narrow game trails, so on the third day, when we sought permission to drink from a Tarahumara family's spring, I inquired about predators. A grizzled elder told us black bears raid their crops and they see tracks of mountain lions and jaguars. The mammals themselves are rarely visible, he added, and, as if by way of explanation, "Esos gatos caminan muy escondidos"—those cats walk really hidden. Asked about rattlesnakes, the old man volunteered only that they're common and bite people, leaving me wondering if his people regard *las cascabeles* as even more inscrutable than felids, and if, like me, they find dangerous snakes charismatic.

During my travels, focused on predators, I've come to believe that nature's most profound lessons, like god and the devil, lurk in hard-won details. As a youth I'd envied George Schaller's landmark studies of Serengeti lions but couldn't conceive of similar research on the smaller, more secretive creatures that captivated me. By the 1980s, though, I was collaborating with Tucson physician David Hardy, and technology made it possible for us to implant tiny radio transmitters in fifty black-tailed rattlesnakes. Over the course of nearly five thousand encounters, we tracked those lovely black and yellow serpents in Arizona's Chiricahua Mountains, chronicling their hunting tactics and spying on their social lives. And throughout that experience, we kept asking ourselves a question that motivates many naturalists: What is it like to be a blacktail, or for that matter a house wren or my dog Riley?

Almost six hundred observations of our most scrutinized rattler began in the fall of 1994, when we located her coiled with male 18 under the leaves of a yucca. While new female 21 was anesthetized Dave detected a meal by palpating her abdomen, and weeks later a substantial midbody bulge indicated she ate again before entering a winter refuge in November. That next March she moved to a rock squirrel's abandoned burrow and in July delivered six

Natural-born killers: black-tailed rattlesnake swallowing a desert cottontail, Chiricahua Mountains, Cochise County, Arizona, July 1, 2003. (Photo: L. M. Chan)

babies. Female 21 remained there ten days, until the youngsters shed their natal skins and dispersed, then crawled forty yards to a white-throated woodrat's nest and hunted for the first time in nine months. She was courted by three males over the next two years, mated with male 27 prior to her 1998 litter, mated with male 26 prior to birthing in 2000, and then skipped three years before her fourth litter.[5]

We especially relished familiarity with individuals. Male 3 courted females but never mated, whereas male 26, at about four feet long and three pounds our largest blacktail, had an enormous range and successfully courted several females. Superfemale 21, as I later called my all-time favorite snake, was an excellent hunter and good mother, plus she stayed out of trouble; over the course of twelve years she showed more meal bulges than any of the others, guarded four litters through their vulnerable first days, and never betrayed her camouflage by rattling. Woodrats and rock squirrels are staple prey for this species, but one morning our star gal struck a desert cottontail, followed the wounded rabbit's chemical trail for more than two hours, and consumed it ninety yards from the ambush site.

Watching the blacktails not only yielded generalizations about their biology but sometimes also left us grinning and shaking our heads in disbelief. One morning male 41 crawled over the cobbles and dry leaves of a shady ravine, stopped abruptly, and for thirteen minutes meticulously tongue-flicked a cliff chipmunk's runway. Then he coiled, his head pointed at the little squirrel's path. Because hunting-site selection had rarely been seen, we lingered, observing with binoculars from a few yards away. A dry fern was centered eight inches into the rattler's strike zone, and after two minutes he extended the crooked neck posture with which males fight over females, crushed the obstructing plant, and re-formed his ambush coil. I shot Dave a skeptical glance and was reassured by his whispered, "He bent down that fern!" Later, after I published those observations,[6] Alberta naturalist Jonathan Wright wrote me of his astonishment at seeing a prairie rattlesnake tamp down grass around rodent burrows before setting up its ambush.

The surprisingly crafty responses of those snakes challenge clichés about minimal intelligence in reptiles, as well as pose questions some researchers believe are unanswerable, even silly: Could male 41 have conceptualized how a plant might thwart his quest for prey, even if the problem manifested itself hours or even days later? Did he employ inferential reasoning and a move usually reserved for vanquishing rivals to solve what experimental psychologists call a barrier problem? What would a naive young male without combat experience have done, and, since rattlers of the opposite sex don't fight, how would superfemale 21 have dealt with that fern? I am among those lucky folks for whom such puzzles keep us headed outdoors, into the lives of others.

Nature has blessed me with many moments when my rumpled soul was naked and yet I felt unafraid. As we walked those cactus-studded ravines, gathering data and imagining the lives of blacktails, I turned from buried grief and self-absorption to more humble notions of our place in the cosmos. Studying predators, I contemplated violence without evil, death without tragedy, as if when their fangs pierced another creature I might accept my own simmering losses. Other memories drift in too, of frogs singing and friends talking softly while high mountain mist enveloped our camp and dusk fell on an African swamp. I recall afternoon shadows in the Mohave and how in that perfect stillness my students were mesmerized by bone fragments protruding from an old owl pellet, then shortly thereafter by the backlit, oversized ears of a kit fox napping by its burrow. Accompanied by kindred souls in such magical places, I would sometimes imagine us lions in the grass, tails twitching, and joy would overwhelm even the most powerful sadness.

The essays that follow address twin themes, the first being how natural historians transform curiosity into science and thereby help save species from extinction. More than that, though, I aim to push into the poetry of field biology, to emphasize the second, more personal theme and explore how nature eases our existential quandaries. I'll begin by introducing the great explorers Charles Darwin and Alfred Russel Wallace, then bring in a venerable institution and another extraordinarily accomplished naturalist, albeit less well known. Since early in the last century Berkeley's Museum of Vertebrate Zoology has played leadership roles in research, education, and conservation, and Henry Fitch, my most influential teenage mentor, got his Ph.D. there in 1937. I've enjoyed a life in some ways parallel to Henry's and for twenty years was employed by the M.V.Z., so Part One combines our stories to illustrate how childhood passions, chance, and opportunity shape adult trajectories.

Part Two moves from youthful obsessions to academic jobs, and thence into deserts and rainforests, looking for snakes and other creatures. We'll get acquainted with the nuts and bolts of field research and teaching, contrast the emotional impact of hot dry places with hot wet ones; we'll learn some basics of serpent biology and examine ways in which fear plays into relationships with limbless reptiles. Part Three begins with reflections on friendship and happiness, then delves into how an eighteenth-century philosopher's aesthetics and Darwin's theory of "descent with modification" can enhance appreciation for biodiversity. We'll also tackle troublesome notions like anthropomorphism and wilderness, and finally, backpacks brimming with questions, hit the trail after answers. My overarching claims are that organisms remain the core of biology, science plays key roles in conservation, and natural history offers an enlightened form of contentment.

With this book I've set out to praise sweeping dry plateaus and soggy tropical floodplains, as well as the black-tailed rattlesnakes, jaguars, and other creatures that enliven them. More privately, though, right from the start, I wanted to thank my heroes and explain some things to friends and loved ones; I knew there would be disturbing undercurrents and thought they'd be straightforward. Instead, *Tracks and Shadows* has unfolded as a complex, rewarding journey during which, after decades of studying predators, I became one myself. Along the way, problems that seemed easy proved intractable. Nonetheless, by sharing my search for solutions, I aim to persuade others to get out there and learn more about themselves. By portraying field biology as art, I hope to add another brief in defense of the wild.

TWO

Naturalist

"DESCENT WITH MODIFICATION," as Charles Darwin succinctly characterized evolution, encapsulates two indisputable facts. First, all organisms share ancestors from whom they have descended. My beloved Riley comes from a long line of Labrador retrievers, but his heritage extends back more than ten thousand years to Eurasian gray wolves. My father's lineage traces to Nathaniel Greene, a Revolutionary War general, and the most recent forebears I have in common with Jesús Sigala, my Mexican Ph.D. student, lived in Europe, many centuries ago. The second fact is that traits like Riley's yellow coat color and the blue-gray hue of my eyes arose as mutations in individuals whose parents lacked them, and those changes were inherited genetically by their offspring.

Darwin's clever phrase thus accounts for similarities (inherited from parents) as well as differences (new traits in offspring) among individuals, and its implication, that strikingly dissimilar organisms share a family tree, extends to all of life. Twisted-tooth narwhals, brightly splotched orcas, and their Eocene kin with small but obvious legs look more like each other than they resemble sharks, reflecting joint heritage as cetaceans rather than cartilaginous fishes. White bats making tents from leaves, western pipistrelles sheltering in crevices, and exquisitely preserved, sixty-million-year-old bat-winged fossils resemble each other more than birds, just as expected if chiropterans diversified from a single furry ancestor rather than a feathered reptile. And mammals as different as whales, bats, and people, whose hairy collective progenitor nursed young more than a hundred million years ago, share more similarities with each other than with sharks or birds.

"Descent with modification" doesn't in itself explain a third observation, that organisms seem designed for particular lifestyles. It doesn't tell us why limbs changed into wings in one lineage and flippers in another, or how chameleons, woodpeckers, and anteaters independently acquired astonishingly long tongues. In 1858, however, Darwin and another Englishman, Alfred Russel Wallace, rocked the literate world by proposing that natural selection drives evolution. Unchecked breeding, they observed, would lead to overpopulation except that many die young; moreover, individuals with particular colors, tooth shapes, and so forth are better suited than others for local conditions. As a result, those individuals are "selected" in the sense of contributing more offspring to future generations, and populations diverge as environments change over time. Overwhelming evidence confirms that common ancestry underlies all biological diversity, and later we shall revisit the extent to which organisms actually are adapted to their surroundings.

Darwin and Wallace came by their insights in the course of direct exposure to nature. Both men were widely read, boundlessly inquisitive, and steeped in Victorian traditions of collecting and identifying organisms. Each puzzled long and hard over the origin of species. The aristocratic Darwin, having passed on careers in the ministry and medicine, used family connections to arrange a five-year trip around the world. His famous voyage on the *Beagle* allowed the young naturalist to view similar geological strata in widely different places, marvel over fossils of extinct creatures, and ponder peculiarities of island life, most famously in the Galápagos Archipelago. He never again went abroad, labored for more than twenty years on "one long argument" for selection, and then anguished over prospects that a lesser-known biologist would steal his thunder.

Wallace could scarcely have been more different, born into the working class and financing his travels primarily by selling specimens to museums and private collectors. As a young man he spent four years in South America, briefly returned to England, and embarked on eight more years of fieldwork that culminated in a classic book, *The Malay Archipelago*.[1] Incessantly curious, unfazed by discomfort and disaster, he conceived of natural selection during a malaria-induced stupor, announced his insight to Darwin in a letter, and always deferred to the older man as chief architect of their theory. In modern parlance, Darwin seems to have been obsessed with reaching the peak while Wallace was at ease on his own path.

As a Missouri teenager with little knowledge of Darwin and none of Wallace, I scoured the countryside for animals. Inspired by Roger Conant's *Field Guide to Reptiles and Amphibians,* eager for wilder times in wilder places, I also daydreamed about particular species that caught my eye.[2] Texas alligator lizards, for example, were first discovered in the Devil's River country and named *Gerrhonotus infernalis* in 1858 by the Smithsonian's Spencer Fullerton Baird, but a century later, when I encountered them in books, there was still almost nothing known about those snaky, bright-eyed creatures.[3] Imagining myself a trailblazing naturalist, I pored over accounts in Hobart Smith's *Handbook of Lizards* of closely related West Coast species—especially quotes from Henry Fitch, who deemed them unusually intelligent and reported a southern alligator lizard holding off three yellow-billed magpies by hissing and threatening with open jaws, tail curled forward like a shield.[4]

Maybe, I thought, I'll roam the Rio Grande borderlands someday and learn something equally exciting about *Gerrhonotus infernalis!*

Fitch was among the "influential saurologists" profiled in *Handbook of Lizards,* and a photo of him caught my attention. The other men (and one woman) were obviously posing, but he wore a straight-brimmed World War I cavalry hat and looked intense, as if distracted from an important task. Publications in our local college library gave a University of Kansas address, so in 1962 I wrote him announcing my upcoming herpetological career and asking about fieldwork that I planned to conduct in Texas. However pretentious my letter, within a couple of weeks back came Henry's cordial, handwritten explanation of how to determine an alligator lizard's sex: "By grasping the base of the tail, twisting it very slightly, and exerting slight pressure with the thumb on one side, ventrally, one usually can cause a hemipenis to be momentarily exposed. Or failing in several such attempts, one may be reasonably sure that the specimen is a female."

I knew that "ventral" referred to underside and "hemipenes" were the sex organs of lizards and snakes, but not that a high school internship with Henry at K.U. would soon set my life's course, or that he would author almost two hundred publications—comprising more than four thousand pages—on plants, snails, spiders, and diverse vertebrates. This unassuming man started graduate work in 1931, when the discoveries of Darwin and Wallace were only decades old, and took his first academic position in 1948, five years before James Watson and Francis Crick unraveled DNA. Decades later, after he summarized half a century of fieldwork at a symposium in his honor, a student would wryly note that she couldn't call only four years of horned lizard research "long term." Applause typically occurs *after* such presentations, but Henry's arrival at the podium that day provoked a standing ovation.

Fitch's life reflects a conundrum with broad implications, one that has dogged my own career. Darwin, Wallace, and countless others have been drawn into nature by orchids, beetles, or whatever seized their fancy. If young naturalists become scientists, however—and this is truer now than ever before—acclaim more likely flows from generalizing than from gathering facts. Ernst Mayr, a prime modern example, is renowned for synthesizing evolutionary theory, not for his discovery of some four hundred new kinds of birds. At first glance, then, Henry's lifelong focus on organisms themselves seems anachronistic, the esteem in which he's held a bit unexpected. In the chronicle of our lives that winds through this book, I set out to illuminate his stature as well as the enduring value of natural history.[5]

Henry Sheldon Fitch was born at the family home in Utica, New York, early on Christmas day of 1909.[6] Chester, his father, graduated from Williams College, and then dropped out of Harvard Medical School in favor of life as an agriculturist. His mother's family had been in Massachusetts since the 1600s, and despite a Boston finishing school background in music and literature, Alice Chenery Fitch enjoyed the outdoors and often took her children on long walks. In 1910, with two young kids in tow, Chester and Alice moved to Oregon and settled on 116 acres of pear and apple orchards, at the southern end of the sparsely populated Rogue River Valley.

The Fitch place had commanding views of the surrounding countryside, and for much of Henry's childhood, regardless of weather, he slept on the screen porch, exposed to the sights and sounds of nighttime. An early fascination with animals was fostered by the family's library of natural history books, conversations with his father, and their rural lifestyle. As a boy he ranged over nearby wildlands, replete with scrub oaks on foothill slopes and heavily forested at higher elevations in the Siskiyous. He fished for trout, shot and trapped California ground squirrels, caught western fence lizards and two species of kingsnakes for pets, and even as a five-year-old impressed onlookers with his fearless handling of the belligerent local gopher snakes.

When Henry was thirteen Chester and Alice sent him and older sister Margaret back to Utica to "absorb some culture," but as he put it, "The effort was a failure." Photos from that trip show a slender teenager with high forehead and thick hair parted a bit to the left, taller than his grandfather and wearing a tie. In one image Henry is holding a northern watersnake, and it couldn't have helped cultural development that his relatives, convinced he'd caught a copperhead, killed the harmless serpent! Nonetheless, the young wanderer became an award-winning ecologist. Like his Berkeley graduate

Henry Fitch, age eleven, with trout, Jackson County, Oregon, 1921. (Photo: A. F. Echelle)

school mentor Joseph Grinnell and other great naturalists before them, Henry was motivated by love of the outdoors and an inborn drive to explore and make sense of his surroundings.

The future professor's academic beginnings weren't promising. Henry attended a one-room country school in which the lone teacher supervised eight grades of rowdy backwoods kids. Owing to a late birthday, he started while still five years old, jumped from second to fourth grade when there weren't other third graders, and was ill prepared for high school in nearby Medford. He graduated nonetheless and entered the University of Oregon at sixteen, only to find a curriculum aimed at pre-med students whose interest in vertebrates was confined to a required cat anatomy course. There were no role models for a zoology career—no one on the faculty could identify a local frog or bird, and even meeting one of his heroes proved disappointing. When he was twenty, while on vacation back in Utica, his aunt Jean arranged a visit with Bronx Zoo curator Raymond Ditmars, whose widely popular reptile books Henry had read. When the young man announced, contrary to what Ditmars had written, that some alligator lizards lay eggs instead of giving live birth, the famous herpetologist huffily dismissed him.

Fitch's failure to excel in coursework at Oregon was no doubt due to boredom and lack of favorable mentors, and perhaps also immaturity and extracurricular activities. He was on the university's cross-country team, ran the mile and two-mile events in track, and owned a motorcycle. His dad bought the wrecked Indian Scout for five dollars and persuaded their handyman to restore it to working condition, after which Henry ranged ever more widely in search of reptiles. He finally took a bad spill in tar and gravel on the way to collect horned lizards near Klamath Falls, and his grandmother, visiting from Massachusetts that summer, prevailed upon him to give up the motorcycle.

As those rocky years drew to a close, paleontologist Earl Packard, who taught field geology at Oregon and knew Henry's father, urged the son to pursue advanced studies at the University of California. Henry already had been impressed by quotes from a 1916 paper on Colorado Desert fauna by Berkeley undergraduate Charles Camp, which he'd read in John Van Denburgh's *Reptiles of Western North America*.[7] Here, finally, was a course of action for an aspiring field worker: observe animals, catch them for study, publish the findings, and thereby contribute to science. By 1931 Camp had received a Ph.D. from Columbia and returned to his alma mater, so Henry, having worked on the family ranch for a year after graduation, headed south. He soon fell under the spell of one of the country's most eminent biologists, a man who by dint of his own rugged background would appreciate an unconventional newcomer from the Rogue River Valley.

In 1908, Annie Alexander, a naturalist, hunter, and heiress of sugar and steamship companies, founded the University of California's Museum of Vertebrate Zoology. Miss Alexander, as all but closest friends knew her, had corresponded with C. Hart Merriam of the U.S. Bureau of Biological Survey about her collection of mammal skins and skulls.[8] Merriam was a distinguished scientist and administrator, having discovered many new species and formalized a theory of ecological "life zones" from studies in Arizona's San Francisco Peak region, and he encouraged Alexander to make specimens the core of a West Coast vertebrate research center. On the basis of correspondence about shared goals and a favorable first meeting, she selected Joseph Grinnell, a young faculty member at Throop Polytechnic Institute (now California Institute of Technology), as M.V.Z.'s first director.

Grinnell was of New England ancestry, like Fitch, but his outdoor skills traced back to a childhood among the Sioux and later experience with indigenous Alaskans. His father was a medical officer for the Pine Ridge Reservation, and Chief Red Cloud, one of Crazy Horse's contemporaries, called the boy "my little friend Joe." Having subsequently prepared bird skins and published scientific papers as a teenager, Grinnell accomplished so much at Throop that he was already well known when he moved to Berkeley. For the next thirty years he studied geographic variation and the origin of species, refined ecological concepts of niche and competitive exclusion, and stood as an early, forceful proponent of conservation. His vision emphasized specimen-based science within an intellectual framework of education and research. Rather than simply building collections, M.V.Z. would synthesize knowledge of organisms, behavior, environments, and evolution—and in testament to their resilience, those principles have guided subsequent researchers into the twenty-first century.[9]

When I reached Berkeley as a new faculty member and curator in 1978, forty years after Grinnell's death, he still dominated local lore with regard to field acumen and scientific standards. One morning in the Mohave Desert, M.V.Z.'s founding director reportedly looked out of a horse-drawn buggy and, motioning toward a string of tracks in the sand, nonchalantly remarked to students that kangaroo rats were breeding. "Between the hind feet," he probably noted tersely, pointing at large dimples left by the nocturnal rodents' sagging testes. More ominous was the story, related by ornithologist Ned Johnson—among the legendary naturalist's academic grandsons—of an assistant who prepared several bird skins but hadn't attached their museum tags before briefly leaving his worktable. Grinnell was obsessed with tying tags *immediately* so that collection data couldn't possibly be confused; thus the young man returned to find his work jumbled in a wastebasket.

Back then rumor had it that only one in eight completed the rigorous M.V.Z. graduate program, and Henry Fitch's inaugural year was perhaps more difficult than most. He'd entered the University of Oregon as a shy teenager, and those four years around older, more urban students had left him withdrawn, even hostile in some social situations. In 1931, with no financial support that first term in Berkeley, he subsisted on a pot of mush for breakfast and twenty-six-cent dinners at a restaurant near campus. On one daunting occasion he was summoned to meet with Charles Kofoid, who chaired the zoology department and feuded with Grinnell over museum policies; the imperious parasitologist remarked with chilling bluntness that "disgraceful" undergraduate grades would doom the newcomer's doctoral hopes. Henry, however, stubbornly got an A for a paper about herpetology in Kofoid's history of biology course and graduated after his grumpy antagonist retired.

Henry also took classes from other Berkeley luminaries, including paleontology, taught by Camp; invertebrate zoology with S. F. Light; and heredity, evolution, and behavior with Samuel Holmes. He generally did well, but mediocre performance was a chronic threat; a few low marks could result in dismissal, and his near-nemesis was the foreign language requirement. In an introductory class, competing with younger students who'd learned pronunciation from their immigrant parents, hard work yielded only the dreaded C. Rather than risk another bad grade, he slogged through a German translation of Darwin's *On the Origin of Species,* thereby gaining the vocabulary and grammar to pass the graduate reading test.

Things soon improved within M.V.Z., where, despite his rural speech and dress, there was respect for Fitch's resilience, dependability, and field skills. Within a year Grinnell judged him "ideally industrious, with marked independence and originality, does things on his own initiative." For Henry, the museum was "for a long time the center of my universe," one that in his memories always retained a "rather magical quality." In 1933, M.V.Z. gained expansive new quarters in the Life Sciences Building, and grad students were active in the intellectual life of the place. They were employed quarter-time as teaching assistants and got by comfortably on thirty dollars a month; they worked in a superb natural history collection, on a campus where snakes, birds, and small mammals were common. Beginning in his third year Henry assisted with the economic vertebrate zoology class for forestry majors, and each spring he helped Grinnell in vertebrate natural history for zoologists, which he remembered late in life as "the best course" he'd ever taken.[10]

M.V.Z. students in the 1930s knew of Darwin and Wallace, but their main intellectual influence was Grinnell. Although the director regarded supervising them as an unrewarding drain on his time, he nonetheless had a knack for bringing out their best scholarship. During Henry's first semester Grinnell met weekly with him and three others, initially presenting them with recent Ph.D. Alden Miller's shrike monograph as an example to which they should aspire. He introduced Miss Alexander as a special person, and his claim that she put up ten mammal skins a day spurred students to work harder (Grinnell himself averaged fourteen). In those early years only Henry studied reptiles, but any sense of isolation was diminished by visits from such well-known herpetologists as Frank Blanchard and Howard Gloyd of the University of Michigan and Laurence Klauber of San Diego Natural History Museum.

Grinnell was self-confident but shy and austere. He led by example of high expectations and quality work and, although an excellent lecturer, suffered fools poorly. He scarcely issued compliments. Students in vertebrate natural history wrote detailed field notes, and Henry first began recording observations while assisting in that course. Outside of class, the director's style was Socratic. Graduate school's central task is original scholarship, reported in a master's thesis or Ph.D. dissertation, and at first Henry assumed he would be assigned a topic. Instead, one day the taciturn professor asked him to draw up a list of potentially interesting projects. Grinnell responded to Henry's first attempt with a question or two, and then told him to return in a week with the list narrowed to the three that seemed the best of the lot, taking into account travel, funding, and other practical aspects as well as scientific interest and prospects for success. Over the course of a month he was further nudged to eliminate all but one. He settled on alligator lizards.

The research entailed collecting specimens, examining stomach contents, taking measurements, and counting scales, then analyzing data and writing up his findings. Local travel was cheap and easy—Henry could catch a ferry across the bay to San Francisco, take the trolley south to a site near Daly City where northern alligator lizards were common, and be back in Berkeley by nightfall. For more distant locales, travel expenses were partly met by collecting gophers for Grinnell. The director sent him out with lists of places from which series were to be obtained and paid him a dollar for each specimen. One hundred and eighty-six trapped, skinned, and carefully documented burrowing rodents financed his first car.

On field trips to northern California and Oregon Henry shipped preserved reptiles back in tin containers and, as was customary, sent postcards relating his progress and requesting supplies. On April 30, 1934, he asked staff researcher Jean Linsdale for "another syringe, since my cracked one broke and slitting the abdomens of my specimens is not satisfactory as the pattern on the skin between scales doesn't show. I am handicapped by cold and rainy weather so haven't filled a tank yet, but I have good series of gartersnakes and other interesting things, including both kinds of kingsnakes." Four days later, Grinnell responded, "Dear Fitch: Linsdale has gone to Nevada. Under separate cover I am sending you a syringe from Ward Russell's room and it's up to you to make peace with him when you get back! More gophers from the Grants Pass country would be exceedingly welcome—also from the higher mountain tops, especially if accompanied by statements of conditions of soil, vegetation, and local temperature."

Of course, all was not science back on campus, and Henry, however shy and low-key, was part of the male-dominated social scene. There weren't many women around M.V.Z., and even Miss Alexander and her long-term companion, Louise Kellogg, were rarely seen; they collected numerous valuable specimens, including bears and other large mammals, but never accompanied men on field trips. Those few women with ambitions faced formidable obstacles, explicit and otherwise, including being mainly regarded as romantic prospects. Later in life Henry would applaud their progress in academia and take pride in his own biologist daughter's accomplishments, but in the 1930s he was a reserved young man. When he found himself sharing a botany class lab table with two coeds, they seemed so uncommonly attractive that he had trouble drawing pinecones. The distractive effects were compounded by their intense dislike for each other; when they spoke, it was only to him, and then only about plants.

The other M.V.Z. men were variously collaborating, feuding, and outright brawling, but overall an atmosphere of friendly, spirited camaraderie prevailed. Among the faculty, E. Raymond Hall, who'd received a Ph.D. with Grinnell and been hired as mammal curator, entertained graduate students with hilarious, dead-on imitations of senior professors and ornithologists of all stripes. Among the students, themselves an eccentric bunch, Seth "Bennie" Benson was especially prone to commotion and conflict. Bennie's dissertation was about the ecological significance of hair color in rodents on white sands and black lava. Once, on a field trip, he'd parked a museum vehicle in a desert arroyo, only to have a flash flood sweep it away, together with his specimens and field notes. On a Nevada expedition Bennie repeatedly drove the truck into sand without reconnoitering his route, which resulted in hours of shoveling instead of collecting, as well as bitter complaints from his companions about such poor judgment. Years later, after he returned to the museum as Hall's successor, Bennie's animosity toward a visiting Canadian researcher who'd made a pass at his wife erupted in a loud fist fight, and he was almost thrown over a rail onto some specimen cases.

Bennie heckled the students too; he was especially critical of their token herpetologist. On the Nevada expedition, Henry spent hours preparing boards as "swamp skis" for crossing mud flats to a small island, only to sink into the treacherous slurry when his invention failed, having to be rescued with a rope. A few days after the embarrassing fiasco, Bennie abruptly asked Fitch to wrestle. Henry was delighted to relieve pent-up frustrations, and they went at it hard in the dirt, cheered on by the others. Henry achieved a ferocious headlock and pinned Bennie so fervently Hall, who'd just arrived to take over leadership of the trip, worried that Fitch would break Bennie's neck or choke him. The heavier man, however, had undisclosed wrestling experience, thanks to which a few minutes later he escaped and sat on Henry's chest. To everyone's surprise, the contest ended with no bad feelings.

Publications are the coin of the realm in science, and Henry's first, a 1933 report on Oregon birds, was followed by three notes in the American Society of Ichthyologists and Herpetologists' journal, *Copeia*.[11] His trademark monographs commenced with two papers from the thesis on alligator lizards, of which one analyzed scales and color patterns, the other habits and habitats. Both were packed with details and had straightforward goals: to set forth the external features and ecology of closely related species, with more general insights woven among empirical findings. He commented on advantages of viviparous reproduction, for example, long before it was a popular research topic: "Eggs left in the ground are exposed to attacks of egg-eating reptiles, mammals, and insects, and to extremes of temperature and danger of desiccation, while those carried by the female probably stand a better chance of developing into independently successful young."[12]

Grinnell was pleased with Henry's progress, though concerned about his idiosyncrasies. As the director wrote to G. K. Noble at New York's American Museum of Natural History, "All who have seen this thesis consider it of more than ordinary value. Fitch is an excellent prospect for becoming a productive herpetologist. He reads widely but is not social in his tendencies. He works by himself, minds his own business, and is not communicative except when drawn out. First impressions upon some people result in underestimation of his talents." About then, Hall urged Henry to take an off-campus class in public speaking. At first the younger man, obliged to give an impromptu speech, could only stand silent for several minutes before making a few awkward remarks, but with practice he gained skills to present talks, as well as some much-needed social poise. Four years later Grinnell's first herpetology student had more than justified his confidence. Henry's achievements, over the course of his lifetime, would far exceed those of "a productive herpetologist."

Settling on a dissertation topic was contentious. Henry first worked on alligator lizard evolution, an extension of his thesis, but then initiated faunal surveys of the Rogue River basin, started comparisons of reptiles living in different habitats, and finally proposed an ambitious gartersnake project. Joseph Grinnell was shocked and indignant at such "irresponsible" vacillation, and after heated arguments they met with Jean Linsdale, who unlike the director had dabbled in herpetology and as a staff member could veto Henry's choice. In front of their boss Linsdale smugly claimed to have suggested Bay Area *Thamnophis* when Fitch first arrived and added, "It was a good project then and still is." Henry had in mind a geographically expansive, evolutionary investigation of an entire species complex, but Linsdale's statement satisfied Grinnell, and so gartersnakes it was.

Alexander Ruthven had noted in his 1908 University of Michigan dissertation that *Thamnophis* "long stood in the minds of herpetologists as synonymous with chaos," and Henry addressed an especially troublesome subset of that mess.[13] Members of Ruthven's "*T. elegans* group" were colorful and abundant, running the gamut from slender desert snakes to stout-bodied savanna inhabitants; in some places scalation and color patterns distinguished two forms, whereas elsewhere variation was kaleidoscopic. Ruthven judged that the group comprised a widely distributed *T. elegans*, *T. ordinoides* of the Pacific Northwest, and *T. hammondii* south into Baja California. Ten years later, California Academy of Sciences' John Van Denburgh and Joseph Slevin revisited this "most difficult problem" and lumped them all as *T. ordinoides*.[14] Like Ruthven, they were hampered by limited exposure to live snakes, since colors fade in preservative, and by a paucity of specimens from critical areas. All three struggled with the questions that inspired Darwin, Wallace, and now Grinnell's student Fitch: Among all those snakes in all of those places, how many species are there, and what accounts for their diversification?

Henry knew from his Oregon childhood that certain gartersnakes were associated with particular habitats and diets. *Thamnophis ordinoides hydrophila* was abundant along rocky streams and ate fish and tadpoles; *T. o. elegans* was scarcer, found in dry oak woods, and fed on lizards and mice; and *T. o. ordinoides* lived in low, wet places but didn't enter water, and subsisted on slugs. By concentrating on areas with two or more "kinds," Henry hoped to determine whether they were geographical races of the same species (also called subspecies) or separate species. He examined almost three thousand specimens, of which he'd personally collected about 850 at 150 localities; he recorded scale counts, head length, and so forth, and for live snakes wrote descriptions of behavior and color patterns. He thus benefited from having far more field experience and data than previous researchers, and as a result he uncovered differences that made sense in terms of ecology.

Grinnell and Miller regarded distinctive populations from different places as separate species (a view that made them, in taxonomic vernacular, "splitters"), whereas some contemporaries ("lumpers") believed that *reasonably* similar organisms likely *could* interbreed and so should be treated as subspecies. Henry accordingly viewed *Thamnophis hammondii* in southern California and *T. digueti* in Baja California as distinct from an otherwise widespread *T. ordinoides;* the latter encompassed four terrestrial and three aquatic subspecies, including two new ones that he named *T. o. hydrophila* ("water lover") and *T. o. gigas,* the Central Valley's watersnake-like giant. Moreover, he described a scenario wherein various pairs among the seven races met and produced intermediate individuals at some places, as predicted by the lumpers, while elsewhere two subspecies abutted with little or no evidence of hybrids, implying speciation. His dissertation referred to the entire complex as an *Artenkreis,* a German word for closely related, interacting species.

Henry also contradicted Ruthven's views on evolution. Despite the Michigan herpetologist's lack of field experience with western gartersnakes, he presaged an anti-adaptationism popularized decades later by Harvard's Stephen Jay Gould, arguing that tendencies to dwarfism rather than natural selection drove diversification. Conversely, Henry's data—he had 945 prey items from the stomachs of 462 individual gartersnakes—suggested that size, shape, and anatomy are related to environmental factors, implying selection. Large forms have stout bodies, more scale rows, unicolored or checkered patterns, small lip scales and salivary glands, and long, narrow heads, all of which might facilitate aquatic habits and a diet of bulky fish. Their more slender terrestrial kin, with fewer scale rows, bright stripes, large lip scales and salivary glands, and short, blunt heads, feed on worms, slugs, and frogs.

After receiving his doctorate in the spring of 1937, Henry worked at M.V.Z.'s new Hastings Natural History Reservation in Carmel Valley while preparing the dissertation for publication. Linsdale was a staff researcher there and notoriously unpleasant, but Henry laughed off the older man's endless criticism for the sake of his guidance in conducting field studies. By then the new Ph.D. had caught the attention of prominent herpetologists, and his professor was unequivocally impressed. Karl Schmidt of Chicago's Field Museum of Natural History called Henry "one of the most promising younger herpetologists," and Grinnell responded that he'd "never known so energetic and effective a collector of cold-blooded vertebrates." Grinnell told the American Museum's G. K. Noble that the *Thamnophis* study was "far and away superior to any others. Fitch pounds along quietly, with extreme industry, entirely on his own initiative. He requires no pressure from seniors to get his work done. His thesis, for example, did not lag toward the end, as is the case of most other graduate students."

Grinnell might have added that most doctoral research doesn't cause much controversy either. Reactions to Henry's 1940 monograph in the University of California Publications in Zoology summarizing his dissertation findings ranged from lavish praise to harsh criticism.[15] Ichthyologist Carl Hubbs, in an *American Naturalist* review, judged it "an outstanding contribution to knowledge of speciation."[16] Writing in *Copeia,* Stanford's George Myers called the work "impressive in care, execution, completeness, and physical bulk" and approved of "the definable criterion of reproductive discontinuity, rather than degree of difference, in distinguishing taxa"—but he regarded *Artenkreis* as "neither necessary nor in good taste" (war with Germany loomed) and decried the absence of statistics.[17] Ernst Mayr, an evolutionary biologist of German descent, claimed in the same *Copeia* issue that the use of *Artenkreis* ran "contrary to all modern practices," and in his influential 1942 volume, *Systematics and the Origin of Species,* concluded that "Fitch's strict adherence to the intergradation principle led to an absurd nomenclature."[18]

Henry, preoccupied with postdoctoral research and military service, delayed responding until 1948. In *Copeia* he credited another M.V.Z. student, Wade Fox, for showing that *Thamnophis ordinoides* is restricted to the Pacific Northwest, which meant that most other western gartersnakes are subspecies of *T. elegans.* He also pointed out that "Mayr's excellent book" regarded as single-species populations some other organisms that overlap without interbreeding, yet rejected similar treatment for gartersnakes.[19] Their disagreement mirrored then-current debates about defining species and problems with inferring relationships from anatomical features, whereas now we rely on breeding experiments and DNA sequences. By 1996 Douglas Rossman had assembled still more data and split the three species into six, meanwhile praising Fitch's "exhaustive study, which few if any have matched and none have surpassed. Particularly noteworthy was the use of nontraditional characters in a highly objective, often quantified, manner."[20]

Henry always regarded the dissertation as his biggest achievement because it connected anatomical evolution with ecological variation, as predicted by Darwin's and Wallace's theory of natural selection. In terms of lasting value, however, accounts of alligator lizards and gartersnakes in Robert Stebbins's *Amphibians and Reptiles of Western North America* and *A Field Guide to Western Reptiles and Amphibians* relied on thousands of specimens that Fitch collected and interpreted.[21] Indeed, without all those quantified, vouchered natural history facts, we couldn't talk effectively about organisms, let alone promote conservation. Henry conducted seventy more years' worth of field studies, and later I'll argue that they've had even greater impact than the early work. But first we'll examine how his practice of natural history inspired others, myself included.

THREE

Nerd

MY MOST CHERISHED POSSESSIONS include an antique kitchen implement, a WWII aviator's memoir, and a rifle almost as old as I am. Walter Hyson Gibson gave the rolling pin to Hattie Lola Crews Gibson in 1921, three months after my mother, their first child, was born. He'd carved his teenage wife's only Christmas present from a single block of oak, the darker heartwood visible down one side; it feels remarkably heavy, off-round enough to dispel any suspicion that Grandpa used a lathe and shiny from decades of Grandmommy's biscuit dough and cobbler crust. My father, the second son of Harry Horace Greene and Bertha Crawford Greene, sent his parents *Bombers Across* to illustrate the travails of his B-24 crew, and only after Daddy's death did I discover in the margins his penciled reflections on the war.[1]

About the same time as I came by the rolling pin and the book, my Berkeley graduate students, knowing that I like Western memorabilia and grew up with a rural appreciation for guns, bought me the model 1894 Winchester as a sabbatical gift. Beyond the aesthetics of its classic lever-action and worn, dark walnut stock, my old carbine epitomizes adventure and self-sufficiency as well as something murkier. Out on a shooting range I enjoy simple pleasures of marksmanship, the mental discipline of lining sights up on a target, letting out half a breath, and *squeezing*, not *pulling*, the trigger; I marvel at the controlled explosion of a .30-30 round, fantasize bygone eras when I would have brought game back to the cooking fire. And sometimes, in that violently concussive moment, time itself seems to shatter.

I was born in September 1945 of a nomadic union, ten months after Daddy returned from overseas. Marjorie Nan Gibson grew up in the East Texas piney woods, her father a WWI veteran and subsistence farmer. Raised without electricity or indoor plumbing, my mother sewed clothes from chicken feed sacks and worked for room and board while attending a nearby junior college. After a year, eager to escape the meager family circumstances and barely nineteen, she took a secretarial job in Washington, D.C., with her district's congressman. Harry William Greene, the younger son of a shoe factory worker who loved books and wrote poetry for the local newspaper, grew up in Endicott, New York. My father, known as Bill, was so enthralled by flight that after a stint with the census bureau he joined the Royal Canadian Air Force, then transferred to the Army Air Corps when we entered the war. Marjorie and Bill met as tenants in a D.C. apartment building, married two years later, and were separated for months while he navigated bombers in Europe, Asia, and Africa. I was named Harry Walter Greene after my grandfathers.

We were never in one home more than three years at a stretch until Daddy retired from the air force, just before I turned seventeen, so over the course of twelve grades I attended schools in five states and a foreign country. Bill and Marjorie were strict but supportive parents, and although our family attended Methodist and Presbyterian churches, my younger brother, Will, and I were encouraged to read and think independently. We were good students and in our spare time flew model airplanes, watched Saturday matinees about cowboys and war heroes, and romped with Sunny, our collie. Mother and Daddy weren't inclined toward the outdoors for its own sake—she owned a dirt farmer's practical view of nature; his hobbies in youth were ice hockey, reading, and photography—but they mustered enthusiasm for countless zoos and museums, tolerated my wild critter pets, and otherwise nurtured our interests. My favorite childhood books were Elizabeth Baker's *Stocky, Boy of West Texas,* about the adventures of an orphaned pioneer kid, and Raymond Ditmars's *Snakes of the World.*[2]

I developed passions for travel and the outdoors at an early age. As a four-year-old, while Daddy worked at nearby Lowry Field, I was mesmerized by dioramas of bison and other large mammals at the Denver Museum of Natural History, and on family outings we used a child's primer on domestic species to identify Hereford bulls and Yorkshire pigs. By age seven I was catching box turtles and horned lizards on visits to my grandparents' East Texas farm, and that same year, on a summer camp hike while we were stationed in the central Texas Hill Country, I watched a western diamond-backed rattlesnake crawl into a cactus patch. Shortly thereafter we left Sunny with the grandparents and took an idyllic troop ship voyage to the Philippines, during which Will and I pretended to be sailors and watched flying fish, whales, and other marine life. For the next sixteen months we lived on the island of Luzon, my roaming tendencies curtailed by cobras and communist insurgents in hills neighboring our base, but finally, when Daddy was assigned back to the same Texas outpost, Sunny and I were again free to explore our surroundings.

After school and on weekends I scampered across the street from our quarters on Gray Air Force Base into scrubby woodlands and shallow lime-stone outcrops, imagining myself an adventurer. I marveled at bluebonnets and other spring wildflowers, always on the lookout for what my mother labeled "varmits" or "creepy crawlies." Evidently neither parent thought of those excursions as hazardous for a nine-year-old. Once, however, I scrambled under huge tree roots on a stream bank, entranced by cobwebs and dank earth smell, only to find the frog I was chasing replaced by a prowling Texas coralsnake. From Ditmars's book I realized the color pattern a few inches beyond my nose fit "red and yellow, kill a fellow" rather than the "red and black, venom lack" of a harmless milksnake, so I knew better than to grab it. Another day I found Sunny lying on our patio with fluid draining from paired punctures on her nose, recognized the wounds as snakebites from my reading, and vowed to henceforth kill every rattler I found. She recovered thanks to good veterinary care, relieving me of a promise I surely couldn't have kept.

Almost every year we visited Mother's parents near the tiny town of Lindale, inspiring my most provocative childhood memories as well as some sense of rooted place, a familiar and permanent home to which we always returned. Grandpa and Grandmommy's weathered plank house was perched on stacked stone pillars, its crawl space open on all sides to chickens and grandchildren. Grandpa drew buckets of cool, tasty well water with a rope and pulley, and everyone bathed in a metal tub on the back porch; there were chamber pots in the bedrooms and a two-seater outhouse, discretely positioned beyond the chicken coop. At meals we kids perched among the grownups on rough-hewn benches, savoring fried chicken and cream gravy, mashed potatoes, snap peas or other garden vegetables, and fresh baked rolls, generously buttered and sopped in molasses. On Sundays we'd climb into a mule-drawn wagon for the mile ride to church, then sit spellbound while the old ladies wailed out "Leaning on the Everlasting

Author, almost six, with sunfish near his grandparents' farm, Smith County, Texas, 1951. (Photo: H. W. Greene)

Arms" and finger-sized katydids crawled from the pews up onto their sunbonnets.

For Will and me, oblivious to the hardships of country life, those Lindale trips were grand adventures. We wore denim overalls like Grandpa's and played among clucking hens, used twigs to noodle ant lions from their sandy pits, and made our own "play pretties," as my grandparents called toys— steamrollers, for example, out of old oil filters dragged around on coat hangers. I awkwardly milked the Jersey cow and watched with rapt curiosity as Grandmommy converted fresh-drawn cream to butter by sloshing it with a pole in a ceramic churn. A fireplace provided heat in winter. We passed one icy morning wide-eyed with questions as a pig was dispatched with a bullet to the head, then bled, scalded, shaved, and sliced up for the smokehouse— "Wow ... so are those the *intestines?*" My mother's family used single-shot rifles and shotguns for putting rabbits, squirrels, and game birds on the table, and we were matter-of-factly taught firearm safety and marksmanship. To this day I would no more assume a gun is unloaded or carelessly swing the muzzle toward another person than step in front of a moving truck.

In 1958 the air force granted our father a year at the University of Oklahoma, and I attended eighth grade at a progressive campus school. Surrounded by the flattest land we'd ever seen, Will and I caught bullsnakes on the nearby prairie, fussed over hairless babies from a road-killed opossum, and launched homemade rockets. My brother had a knack for sports, and by dint of genes, upbringing, or both he shared Daddy's talents for fixing anything from cameras to automobiles, whereas I was clumsy, worthless at athletics, and utterly lacking in technical skills. We grew up liking books, and I relished a sense of *writing* something—one class paper was about Mexican axolotls, relatives of tiger salamanders, and I tried to imagine actually discovering all those facts. Popular music captivated us too, of which the Kingston Trio's mournful ballad about Tom Dooley, hanged for stabbing his girlfriend, still comes to mind. Never having heard a cross word between my parents, I couldn't imagine a man killing the woman he loved, let alone one day knowing such people.

I look back on my teenage years with gratitude for the influence of several zoologists, beginning at O.U. One day, Daddy came home with news of a professor who studied herpetology, so I biked over and introduced myself. Charles Carpenter showed me lizard nooses, snake sticks, and jars of specimens collected on his expeditions. He talked about careers in biology, then recommended James Oliver's *Natural History of North American Amphibians and Reptiles* and Roger Conant's *Field Guide to Reptiles and Amphibians*.[3] That night over dinner I announced that I too would become a professor, and my folks soon gave me both books as thirteenth birthday presents. Oliver laid out lifestyle details not covered in older, more general works, and I was surprised that Conant's small volume encompassed some three hundred species. On a class trip to O.U.'s biological field station, I was even more impressed by Carpenter's outdoor enclosures for observing lizards and his skill at capturing a snappy coachwhip snake.

Thanks to Daddy's next assignment we lived two blocks from Central Missouri State College in Warrensburg, and that spring I accompanied a college class canoeing and caving in the Ozarks. On our first afternoon out, while I drank from a rivulet, a student clambering upslope dislodged a baseball-sized rock. Seconds later I fell to the side as blood streamed over my face and hands; then the nasty scalp cut was stanched with a bandana and I was ordered to take it easy. The following day things took an upswing when I checked out side passages in Inca Cave and spotted a grotto salamander in the beam of my headlamp. *Typhlotriton spelaeus* is blind and pigment free, and because it wouldn't have visual predators, I was surprised that the three-inch-long troglodyte waved its elevated tail back and forth when picked up. On the ride home I recalled Dr. Carpenter's paper describing similar behavior in surface-dwelling tiger salamanders. A few months later my first publication, a nine-sentence report on the vestigial defensive display of *T. spelaeus,* appeared in the *Bulletin of the Philadelphia Herpetological Society.*[4]

Small-town teenagers who publish in scientific journals aren't likely to excel in social skills. I was a lowly high school freshman when we moved to Missouri, with unstylish short hair and no athletic talents, so I hoped to break the ice by showing my classmates some animal behavior. Having gained our science teacher's bewildered permission, I brought a gartersnake and a leopard frog to school, then passed several hours imagining myself on the cusp of popularity. Early in the afternoon's first period I placed the two in a terrarium on Mr. Smith's desk; immediately, the brightly striped serpent began swallowing its struggling prey by a hind leg. The doomed amphibian let loose a series of blood-curdling screams that were finally silenced by ingestion, and as snake jaws opened and closed over the frog's snout a girl near the front of the class launched us into bedlam. She wailed and flailed her arms, vomited on several other students, and after a couple of minutes fled the room, her cries echoing down the hallway. Then, while the other students settled down, I brooded on the likelihood of never making any friends.

As a sophomore I grew taller, and Daddy bought a razor for my nascent moustache. However socially marginal, I suffered a crush on a shy farm girl in citizenship class. Claudia's desk was in front of mine and I was overwhelmed by the briefest of smiles or the passing fragrance of her hair; I strained for a glimpse of panty line, wondered what it might be like to kiss her lips, and was generally clueless about my fifteen-year-old body. Then one day we gave group reports, lined up by seating order such that I was the last presenter from my row. When Mr. West called out, "Claudia, you're next . . . ," she glanced up, blinked a couple of times, and crumpled to the floor. As the other kids gasped and fidgeted, I remembered Boy Scout training and took command. Grabbing the girl of my daydreams by the ankles, I lifted Claudia's feet to waist level and was stunned when Mr. West shouted, "Stop looking up her dress!" Naturally that prospect would have sent my teenage head spinning, but all that had come to mind was one line in a first aid pamphlet: "For fainting, elevate the legs."

Mostly I reveled in natural history. That next summer I audited a C.M.S.C. class in which we trapped and prepared rodents as museum specimens; sixteenth birthday presents included an album of frog calls and a special India-ink pen for field notes. I joined the American Society of Ichthyologists and Herpetologists and corresponded with *Field Guide* author Roger Conant, who introduced me to Kansas City naturalist Paul Anderson;[5] following their suggestions, I began writing to other biologists and amassing reprints of their scientific papers. A slender glass lizard, dead on a trail in nearby Knobnoster State Park, was cataloged as "HWG 1" for my newly inaugurated collection of preserved specimens. Next-door neighbors Stanley and Mildred Fisher and their kids shared these interests. One weekend we exhibited live and preserved herps in our garage, charging admission, and were featured in the local newspaper. Coal skinks and red-bellied snakes hadn't previously been reported from north of the Ozarks, so Mike Wakeman and I published in *Herpetologica* on those we caught,[6] and with another classmate, aspiring physician Mike Wyatt, I dissected road-killed cats and stillborn lambs from his dad's farm.

All that nerdiness aside, I often joined my friends for fountain cherry Cokes and French fries at Warrensburg's only drugstore, where we envied the guys with letter jackets who were always surrounded by girls.

Despite the gartersnake debacle, our science teacher arranged for me to join other rural Midwest high school students at the University of Kansas Summer Science and Math Camp following my junior year. For two weeks in Lawrence we attended classes ranging from astronomy to zoology, and most importantly, we *did things,* like making acetate prints of cross-sectioned fossil plants and examining live chick embryos under a microscope. Meanwhile my father retired from the air force to teach college in Fort Worth, and after the K.U. stint I joined our family there. By then intent on being a scientist and eager to meet like-minded folks, I attended meetings of a local amateur naturalists' group. My new acquaintances included Ben Dial, who was a year older and like me kept snakes in his bedroom, had published scientific papers, and loved folk music. We traveled all over Texas in his 1957 Chevy, searching for herps, our conversations running the gamut from Texas alligator lizards to the long-haired woman in a folk singing group called Peter, Paul, and Mary. Ben and I would remain friends for the rest of his life.

I was selected with nineteen other Science and Math Camp students to apprentice with a professor and spent the summer after graduation working at K.U. for Henry Fitch, with whom I'd already corresponded. Because camp policy against riding in cars prohibited me from working at the Natural History Reservation a few miles north of town, Henry suggested I study lizard reproduction by examining specimens in the Museum of Natural History, under daily supervision of two graduate students. Jay Cole came with undergraduate research experience from summers at the American Museum of Natural History's Southwestern Research Station in Arizona and was preparing his first papers for publication. Linda Trueb was newly arrived from Berkeley, where she'd been inspired by a natural history of the vertebrates course taught by renowned Museum of Vertebrate Zoology herpetologist Robert Stebbins.

Those two months hooked me on biology. Each day I walked from my campus dormitory past imposing stone buildings that mirrored childhood notions of a university. At the museum I removed skinks from jars of alcohol, laid them out on a tray, and recorded collecting dates and localities from paper tags attached to their hind legs. One by one I slit scaly bellies with sharp-pointed scissors, then counted and measured developing eggs with calipers. I was gathering data, and the lizards provided snapshots—two shelled eggs in each oviduct would have been laid within days of when a female was caught, for example, while five enlarged yellow follicles in her ovaries portended a second, larger clutch later in the season. Collectively my records would describe reproduction throughout a species' geographic range. Our goal was to compare tropical and temperate skinks for a better understanding of how climate fine-tunes breeding.

Henry and I examined more than six hundred specimens and discovered that larger ground skinks laid more eggs than did smaller ones. Moreover, in the southern United States, between March and August, they deposited four clutches of two or three eggs each, whereas northern lizards had time to produce only two clutches of five to seven eggs—so the annual total per female was about the same at geographic extremes. Central American brown forest skinks contained eggs during most months, evidence of a more extended breeding season than in ground skinks—as we expected for reptiles in tropical regions with year-round warm temperatures.[7]

Dissecting pickled lizards was tedious, and I welcomed diversions. One day Jay and Linda assigned me to review literature on reproductive isolating mechanisms in frogs. Sifting through decades of the *Zoological Record*'s annual, worldwide survey of publications, I learned that countless people do herpetology. More importantly, my supervisors introduced me to criticism unlike anything I'd experienced in high school, delivered in red ink all over my report ("The word 'data' is *always* plural!"). Henry's daughter Alice, two years younger than me and studying temperature effects on skink eggs, was also a distraction. Once I followed her into the museum's climate chamber, came back a few minutes later, and found my data sheet marked in Linda's elegant script: "Before you run off with a pretty girl put the cap back on your pen!" As summer ended I got up the nerve to invite Alice to a party—but after we'd had punch and danced a few times, Henry retrieved her so fast that years later I teasingly asked her husband, Tony Echelle, how they'd ever managed any privacy for courtship. "Henry," he answered, "was in Costa Rica that year."

Graduate student life looked a lot more exciting than high school. Jay was gathering material in the Kansas Flint Hills for his thesis, and once a week, before dawn and under threat of expulsion for riding in a car, I slipped down to his apartment and hunkered out of sight as we left town. He hoped to discover the function of the waxy "femoral pores" on the hind legs of collared lizards by analyzing how they changed seasonally, so each trip we collected a male, female, and juvenile for lab work.[8] I also wanted to find venomous snakes, but because our grassland site seemed too far west for copperheads and too far east for prairie rattlers, I wasn't optimistic. Nor, as a result, was I initially cautious. One morning, though, I spotted an old plank that looked just right for reptiles, turned it over, and was electrified by urgent buzzing near my fingers. I hadn't realized the Flint Hills are prime habitat for massasaugas. Fortunately that little rattler didn't bite me, nor, later, did two others of her species, heard rattling in the dew-soaked grass before we saw them.

I found more venomous snakes after returning to Texas, one of which taught me a long-lasting lesson. On one occasion along a creek near my grandparents' place I encountered two thick-bodied cottonmouths; as I'd seen illustrated in books, I pressed each one's head down with a snake hook, then held it between a thumb and two fingers. My heart was racing, the air thick with musk from the struggling serpents, but I got them into bags and preserved without incident. A few weeks later, though, I found a broad-banded copperhead under a stump and took it back to my home laboratory.

Anxious to measure the snake and emboldened by having caught the cottonmouths, I encircled its neck with left thumb and forefinger, stretched its tail toward the twenty-four-inch mark on a yardstick—and felt a hot wasp sting–like jab. I was shocked to see fangs spider-walking toward my right thumb, necessitating rapid maneuvers to avoid a second bite! Fearing adverse parental reactions, and confident from reading about Henry's untreated bite by a larger Osage copperhead in Kansas,[9] I kept the accident to myself and suffered only nausea and mild swelling.

Among my friends in the Fort Worth Children's Museum's natural history club was George Oliver, a couple of years younger and as obsessed with mammals as I was with herps. We'd read in Klauber's *Rattlesnakes* that little was known about massasaugas,[10] so, having commonly found them crossing roads through the prairie west of town, George and I capitalized on the museum's specimens to augment our field observations. On weekends we examined preserved snakes, using the collection data to assess seasonal activity and stomach contents to study natural diet. I could recognize all the local frogs, lizards, and snakes from just a foot or a piece of tail, while George identified a shrew and several species of mice, sometimes based on only fur and a few teeth. Our goal, right from the start, was to publish a scientific paper.[11]

Southwestern University enticed me with a scholarship, high academic standards, and a friendly atmosphere. Georgetown was thirty miles north of Austin, straddling the eastern edge of the Hill Country (formally known as the Edwards Plateau) and thus home to unusually high herp diversity.[12] The town nestled between branches of the San Gabriel River, its physical presence dominated by historic flagstone buildings and church spires; land to the west was devoted to ranching, while to the east family farms covered what was once prairie. Though geographically well positioned for my biological interests, I was also starved for social acceptance and emotionally unprepared for independent living. Freshmen quickly discovered that fraternities and sororities dominated campus life, and of the eight hundred students those few who didn't join were ostracized as "G.D.I.s" (goddamned independents). Within weeks I was sitting on the edge of my dorm bed wondering about pledging a fraternity when a senior ran through the hall shouting that President Kennedy had been assassinated.

A high school neighbor from Fort Worth also enrolled at S.U., and we often went to movies and concerts together. Marsha was self-reliant, bright, and intensely vivacious. Many years later her mother told me of her daughter sitting curbside and copying the license plates of passing cars so she could practice addition and subtraction—"with *real* numbers, Mom!"—and a mutual friend back home named "grace," "integrity," and other such qualities as she recalled idolizing the older girl. When we met, Marsha was a petite seventeen-year-old with green eyes and short brown hair, and I was enthralled by her adventuresome spirit and girl-next-door charms. I was also naive in matters of the heart, and that fall first learned how taste could merge with smell, vision with touch, as our physical attraction grew into emotional involvement. Sometimes my head threatened to explode from her kisses and her shy caress.

I cannot recall now why we drifted apart—only that sororities inexplicably passed over Marsha, and at the end of spring semester, hurt by prospects of pariah status on our little campus, she transferred to a larger, faraway university. Although we saw each other the next couple of summers and once shared an idyllic lakeshore weekend with her family, neither of us ever said "I love you." Maybe we were reserving those words for that special person with whom we'd supposedly spend a lifetime, and in hindsight perhaps one or both of us lacked commitment. Over the years we occasionally got together back in Fort Worth, but because we never endured the conflicts and adjustments that arise between seasoned lovers, my memories of her are too perfect and sadly shallow. By the time we graduated I was dating the woman who would become my first wife, while Marsha had married an oddly quiet guy named Lyndy.

At Southwestern I finally blossomed socially. I began drinking and smoking, and on weekends my fraternity brothers and I drove to Austin for entertainment. We frequented a coffee house, The Eleventh Door, where I was infatuated with the stunning looks and clear voice of Carolyn Hester, who'd preceded Joan Baez and Bob Dylan in the East Coast folk scene before returning to her native Texas.[13] Between songs Hester spoke so softly one strained to hear, yet we hung on every word when she talked about Kennedy's death and sang her sorrowful rendition of Walt Whitman's "Oh Captain! My captain!" School nights were spent at a Georgetown diner, and late one evening, true to the "Lice and Mice" café's reputation, a cockroach dropped from the ceiling into my chili bowl. Because class work held little appeal in the face of such diversions, soon my academic career was in free fall. I failed French, English, math, and organic chemistry; in a calculus exam, my pencil skidded across the page when I conked out midway through an equation, exhausted from cramming weeks' worth of assignments into a single sleepless night.

At the end of my sophomore year, in debt from an ever more expensive social life, I sought employment at a local funeral parlor. My job involved manning an ambulance, driving the hearse or flower car at funerals, assisting with embalming, and dressing bodies, for which I earned lodging and sixty dollars a month. Georgetown had two mortuaries but not enough deaths to make either lucrative. At first I worked for Clarice, who'd inherited the poorer establishment when her husband died in a plane crash. Mr. Armpy's fancier facility was across the street, the hostility between the two families so great that his kids spied on us from a cardboard box in their alley. Nonetheless, he employed me when Clarice sold out and left the area.

Life in a small-town funeral home resembled the wackier scenes in John Irving's *Hotel New Hampshire*.[14] Clarice lived upstairs with her mortician lover, George, and two kids. My tiny apartment was under their stairwell and directly across from the embalming room. Although dead bodies made for a spooky ambiance, wailing mourners rarely disturbed me, and the rambling old house was usually tranquil. One evening, however, I was interrupted in my reading by scrambling claws overhead, followed by high-pitched barking and people hollering. The family's poodle was in heat, and as I emerged into the hallway a large mutt hurtled over the railing, miraculously landed on its feet, and exited through a side door that had been left open when we returned from an emergency call. Within seconds George came running down in undershorts and T-shirt, clutching a shotgun, and disappeared into the back-yard. The 16-gauge boomed once, George returned upstairs without a word, and soon I heard him calling police about shooting a stray dog.

Our all-purpose vehicle was a black Cadillac hearse outfitted with siren and red light, oxygen, a small first aid kit, and two gurneys. For funerals we unscrewed the flasher, removed the medical gear, and flipped up floor rollers so that a casket could slide in and out. The ambulance runs mostly moved patients between hospitals and nursing homes; real emergencies were uncommon, but Williamson County's senior citizens passed away with some frequency, and my first encounter with a dead person came hours after being hired. We pushed the gurney up parallel to the elderly lady's bed and, as instructed, I lifted and pulled from under her knees as George moved her shoulders. The woman's skin was cool and slack against my hands, and—just as I held a deep breath and he elevated her torso—air escaped from her throat with a loud "aaaaahhhhh"!

The mortuary approach to emergency medical care was appallingly simple and uninformed. My Boy Scout first aid training having been deemed sufficient, our main response to injuries and illness was high-speed transport to the nearest hospital. We routinely pushed that big Cadillac past a hundred miles an hour on highways, and when short-handed I ran calls alone, enlisting relatives and passersby to help load patients. Sometimes we rolled far enough south for wrecks on the interstate that an Austin ambulance reached the scene too, setting up a competitive scramble; once I scored a body for us by running with a gurney faster than the other crew. Luckily everyone we carried either was hopelessly moribund or survived in spite of these inadequacies.

One July dawn I crept north out of Georgetown on I-35, hugging the steering wheel and peering through dense fog a few feet in front of the ambulance. I had driven some dozen miles, queasy with anxiety, when several wrecked vehicles emerged around me in the mist. Among them was a tractor-trailer and behind it a Chevrolet Malibu. The car had plowed under the truck with enough force to shear off its top and windshield, then been hit by a pickup and jarred free. Moving gingerly among disjunct bumpers and mangled fenders, over glittering shards of glass and metal, I yelled, "Hey, someone help me!" A woman's panicked voice pierced the chaos: "I can't see . . . what should I do?" The Chevy's left front door was jammed, so I leaned in and checked an unconscious young man slumped against the dash, his nose sliced off and face streaming blood. I got close enough to hear breathing and felt a pulse in his neck, then climbed over what had been the rear window and into the car.

A teenage sibling was slouched into the right backseat corner such that his head seemed tucked out of sight, so I pulled the kid upright by a wrist and discovered his jaw still attached to a ragged neck stump, as was the empty back of his skull. Everything else was splattered over the trunk of the car, the largest bloody gray pieces of brain, bone, and flesh no bigger than pocket change. I groaned and took a deep breath, then lowered him back down, climbed out of the car, and convinced the shaken truck driver to help me get the other boy onto a gurney. That afternoon people gawked with curiosity and revulsion at the wrecked Malibu, which had been towed into town, and twice that next week I awoke to the same nightmare. In my dream the foggy scene played out in colorful detail, except that the shattered young man was my brother, Will.

Another morning that summer a sheriff summoned me to follow him west out of town. We wound our way through several miles of brush country, a rolling jumble of rocks and prickly pear mostly used for ranching and deer hunting leases. Eventually we turned off at a gate, were met by an elderly man in a pickup, and caravanned several hundred yards down a dirt track. The battered, faded blue Oldsmobile was barely visible from the road, tucked down a gully, and in the backseat, as if she'd only fallen over asleep, was a middle-aged woman with short curly hair. A piece of garden hose extended from the tailpipe forward through the front left vent; the windows were closed. Only days earlier this lady had cheerfully sold me snacks and tooth-paste at a convenience mart in nearby Round Rock; now she lay on tattered upholstery, arms extended awkwardly, hands dangling. Her blouse was unbuttoned and bra loosened, shorts and underwear pulled down onto her calves.

After we'd opened all the doors and I touched the woman's tepid neck to make sure there was no pulse, the sheriff picked up an envelope from the floorboard. He read the letter without a word, passed it to me, then looked away while lighting a smoke. Two pages of rambling, apologetic scrawl related how Evelyn couldn't live without a guy who also worked at the store, how she loved him desperately, and finally, "Oh sweet Jesus, please forgive me." I gazed at her vacant eyes and remembered "Darrel" stitched on the shirt of a tall fellow who wore a cowboy hat and said little beyond that required for selling groceries. The old rancher read the note and pursed his lips, where-upon the lawman pointed to a ring on the dead woman's left hand and said, "Let's keep this quiet . . ." We nodded and went about the business of loading her up.

Williamson County didn't have a morgue, so with the sheriff watching, a visiting medical examiner performed an autopsy in our funeral home. He deftly opened the woman with long scalpel cuts, beginning with a shoulder-to-shoulder half moon across her chest and an extension up to it from below the navel. After another incision in front of one ear, down around the base of the head, and up over the other ear, he pulled her scalp forward to just above the eyebrows, laid it over her face, and exposed the cranium. A few seconds' work with a small power saw freed the skullcap, after which, using both hands, the pathologist lifted her brain up for inspection. Next he carved out a shieldlike chunk of ribs and chest muscles, exposing internal organs. With the nonchalance of a store clerk sorting vegetables, he palpated heart, lungs, liver, stomach, ovaries, and so forth for gross abnormalities, then cut off small samples of tissue for study.

The sheriff was sweating profusely but said little beyond an occasional "Oh my lord . . ." and "Huh, well . . . would you look at that." I'd told them I was majoring in biology, and the medical examiner tersely answered my ques-tions about anatomy. He noted a pinkish cast throughout the body and told us that was typical of oxygen-starved tissues; the two men seemed satisfied that the cause of death was carbon monoxide poisoning. No one mentioned the state of the woman's clothing or the note. As they left, the funeral director asked me to sew her up. I was almost twenty-one years old, surprised that human skin was so tough but even more incredulous that such things happened to ordinary-looking people in small Texas towns. I had no idea how much worse life could get.

FOUR

Field Biologist

MY PARENTS' GENERATION EXPERIENCED the 1940s as itinerant, uncertain, sometimes dangerous years. For Henry Fitch the decade following graduate work encompassed a burst of research as well as marriage and finally a move to Kansas, events that proved central to his happiness. First, however, the newly minted Ph.D. needed a job. Although one West Coast college expressed interest, Henry favored fieldwork and thus joined a group founded by Annie Alexander's friend C. Hart Merriam. By then the Bureau of Biological Survey was focused on pest control, so the new Ph.D.'s first task was determining the impact of rabbits and rodents on ranching in the Sierra Nevada foothills. On March 31, 1938, he wrote Joseph Grinnell from the San Joaquin Experimental Range that "associates and surroundings here are agreeable, but I have difficulty confining my attention to ground squirrels amid the abundance of reptile life."

Any naturalist knows that snakes, raptors, and carnivores eat herbivores, so Henry soon set his sights on studying them all, but the Range superintendent disapproved and assigned someone to monitor his new employee. With characteristic tenacity, Henry "bootlegged" the predator work and over the next five years, his boss's obstructionism notwithstanding, gained phenomenally well-documented insights. Kangaroo rat diets were assessed by sorting 5,371 seeds from cheek pouches, sifting 37,904 seed hulls from burrow mounds, and recovering 5,024 items from burrow caches; 9,135 prey items were identified in the droppings and stomach contents of eleven species of predators![1] World War II, however, delayed publication of the results for almost as long as had been required to compile them.

Henry was drafted on four days' notice in the spring of 1941, at age thirty-two. The "influential saurologist" portrait that later caught my eye was taken during his first few months in the army. Henry trained as a medic and pharmacist at William Beaumont General Hospital in El Paso, and in deference to family pride—his paternal great-grandfather also was a pharmacist—he sent home a photograph of himself wearing the only headgear available in the military studio. In September a new directive indefinitely extended service only for draftees twenty-six or younger, and he was unexpectedly discharged. Henry gladly resumed research at the Range. Within a few months, however, soon after the Japanese bombed Pearl Harbor, he was abruptly recalled, and this time he was in for a long haul. When Chester and Alice Fitch were asked to provide a photograph for *Handbook of Lizards* while their son was overseas, they sent the one of him in a cavalry hat.[2]

Military life can be brutally anonymous yet highly personal, as I would also learn, and some aspects of Henry's experience are best described in his own words. "In late 1942 the Victory ship *Kokomo* carried five thousand of us across the stormy North Atlantic in quarters that were crowded, dirty, and reeked of sick odors because of lack of adequate ventilation. We were part of a large convoy and had destroyer escort. The hold was partitioned into many compartments that could be used as air space to keep the ship afloat after torpedo strikes, and the unannounced and frequent testing of the electronically operated sliding doors that sealed our compartments was a grim and persistent reminder of the expendability of the individual in times of war." His medical unit was stationed in Wales, then after D-Day moved to a Scottish base to attend to combat casualties from the continent. "We were often routed out of bed during the middle of the night to carry the wounded into our hospital or transfer them onto another plane or train."

Boredom is an inevitable challenge of soldiering, and whenever possible Henry indulged his passion for zoology. During training breaks in Texas he collected Chihuahua Desert reptiles for M.V.Z. On the way to Europe, as his vessel "wallowed through stormy seas of the North Atlantic, auks suddenly broke through the side of a wave, sleek and dripping, and skittered off over the ocean surface, their narrow wings and shallow wingbeats seemingly barely adequate to keep them out of the water." Back on land there were plenty of distractions for a naturalist happily out from under the deck of a troopship: "Along the coast of Wales were large colonies of nesting seabirds on a long rocky peninsula, a great place for bird watching. On weekend trips to London I frequented the zoo, British Museum, and many bookstores. The infamous 'buzz bombs' were a hazard and I heard many explosions, but none was ever very close." Henry passed his last months overseas in France and

Germany, where he delighted in seeing animals like the slow worm, a snake-like relative of alligator lizards.

In November 1945, after almost five years in the army, Henry was discharged at Camp Beale, California, and he passed his first night of freedom hitchhiking to the family ranch in Oregon. Right away he attempted to regain a position at the San Joaquin Range—draftees were guaranteed their previous jobs by presidential decree—but was initially frustrated. The Bureau of Biological Survey had been renamed Fish and Wildlife Service, and Henry was told that although his job no longer existed, the Soil Conservation Service could employ him for vertebrate surveys in the central United States. "Considerably irked" at the government for breaking a promise to soldiers, he declined and threatened to return on his own to the Range for whatever could be salvaged. Within weeks, funds were "mysteriously found" and he was reassigned to the California study site.

That first night after Henry arrived back in Medford there was a party for soldiers at the Rogue River Inn, and his sister-in-law arranged a blind date. Virginia Ruby Preston, vivacious and eleven years his junior, had moved from Los Angeles in high school and earlier dated his brother, Chester. Henry noticed that she'd "matured very pleasingly," and his fatigue from the trip and "spastic dancing style" notwithstanding, they had a magical evening. He soon returned to work in California and, despite parental disapproval on both sides, kept in touch with Virginia by mail and weekend meetings. They married in Nevada on September 6, 1946, writing field notes and reading page proofs on their honeymoon, and set up housekeeping in a little place near the Range headquarters. Henry was almost thirty-seven years old, with a trim, youthful physique and thinning hair combed straight back without a part; photographed a year later on a picnic, sitting in the grass beside his wife and infant son, John, he looked happy.

Henry's research at the Range eventually yielded lengthy publications on California ground squirrels, Tulare kangaroo rats, desert cottontails, red-tailed hawks, and western rattlesnakes, as well as shorter accounts on other vertebrates, totaling fifteen in scientific journals and three in livestock periodicals. Among them, a 1949 analysis of snake populations set new standards for reptile field studies, and he won the Ecological Society of America's prestigious George Mercer Award in 1950 for an eighty-four-page treatise in *American Midland Naturalist* about ground squirrels on grazing lands[3]—made even more satisfying because that manuscript had been rejected by the society's own Ecological Monographs.

Work went well, but dissension between supervisors in the Fish and Wildlife Service and the Forest Service led to Henry's being transferred. In May 1947 the Fitches journeyed to a housing project in Leesville, Louisiana, where he would work on bobwhites, mourning doves, nine-banded armadillos, cotton rats, and white-tailed deer in a National Forest. Virginia was pregnant with Alice and used their personal car for the trip, while Henry drove the government vehicle with little John at his side. Within a year he'd finished an ecological study of armadillos; he began cruising roads to monitor snake activity, but that research ended with one last major move. Thanks to an earlier professional association, the family would settle close to the geographical center of North America, about halfway between Utica, New York, and Medford, Oregon.

E. Raymond Hall was one of Joseph Grinnell's earliest Ph.D.s and so accomplished a mammalogist that M.V.Z. retained him on staff. Grinnell died unexpectedly in 1939, however, and a few years later, when former fellow student Alden Miller took the helm, Hall left in a huff to establish a similar program at his undergraduate alma mater. As director of the University of Kansas Museum of Natural History, he established M.V.Z.-like traditions of specimen-based research, careful attention to field notes, and graduate education, as well as a personal reputation for eccentricities. Hall enforced style preferences on all museum publications, insisting, for example, that nouns never modify other nouns. One was to write "mounds of gophers" rather than "gopher mounds" for the tillage of burrowing rodents, which some colleagues sarcastically said implied heaps of carcasses!

K.U. had owned a one-square-mile tract northeast of Lawrence since 1910, and Hall persuaded the chancellor to allocate it for teaching and research. Henry had first met the director at M.V.Z. and enjoyed a Nevada summer course with him. Hall was impressed with the younger man too, especially on field trips, and familiar with his Hastings and San Joaquin Range work— he considered him a "competent, effective teacher and good lecturer" and noted that "students in the laboratory quickly recognize he is possessed of exceptional information and industry." Hall was even more effusive in terms of research, referring to Henry as "erudite, honest, exceptionally productive, and excellent in the field," and in the summer of 1948 hired him as instructor in ecology and superintendent of the new reserve. After living in campus housing for almost two years, in March 1950 the Fitches moved into a newly constructed residence on the Natural History Reservation.

Henry immediately launched an ambitious research program, one that reflected M.V.Z. training and subsequent field experience, by focusing on many local species rather than a few close relatives throughout their ranges— as he had done with alligator lizards, for example, in California. The Reservation, composed of woodland interspersed with grassland and weedy fields, boasted three species of fish, ten species of amphibians, twenty-four species of reptiles, twenty-nine species of mammals, and 167 species of birds. Hall was anxious to have mammals studied, so Henry concentrated on them, but he couldn't ignore other vertebrates. Abundant and easily encountered species were given priority, including prairie voles, eastern woodrats, Virginia opossums, five-lined skinks, and certain birds.

Henry especially sought recaptures for data on growth, movements, home ranges, and longevity. He divided the property into eighty areas bounded by creeks, outcrops, roads, fences, and so forth, with names like Rat Woods and Quarry Field. Trees, boulders, and other landmarks were sited on aerial photographs and maps, then numbered or named as points from which captures could be referenced. Various live traps totaled in the hundreds, and Henry invented a cylinder with funnels in each end for snakes that is still the industry standard.[4] Rock ledges, other natural barriers, and drift fences directed animals toward the traps, thus increasing their efficacy. He marked birds with streamers and conventional leg bands; most herps and mammals had one or more toes docked in a standard numbering scheme, while snakes had belly scales clipped in unique combinations.

New discoveries, unexpected opportunities, and technical innovations drove shifts in research during those early years. While examining the droppings of five-lined skinks Henry came across glittering green, jewel-like objects that proved to be the chelicerae, or mouthparts, of black and white jumping spiders, so despite mild arachnophobia he undertook an extensive study of them.[5] During an investigation of reproduction in snakes, he probed their vents with grass stems for the presence of hemipenes in males and microscopically examined liquid from the vents of females for motile sperm.[6] Although emphasis was on activity patterns, he also studied feeding ecology by palpating prey from stomachs and using reference specimens to identify dietary remnants in scats.

Checking traps, marking animals, and recording observations under the Kansas sun was brutally tedious. Then there were the chiggers and thunderstorms, not to mention the wasp nests under rocks he lifted looking for reptiles—but Henry just wiped his brow, kept an eye on the weather, and swatted insects. He accustomed himself to thirst, forsaking a canteen for other gear, and by morning's end, drenched in sweat, he'd come in for the gallon jar of iced tea that Virginia always had in the fridge. Body searches for ticks were a family custom. He did office work in the afternoons, when summer heat was unbearable. Occasional surprises broke the routine too, such as the first copperhead lacking hourglass-shaped markings among hundreds he'd examined. During that first decade he published several papers a year, including a thick 1958 volume on home ranges and movements of 233 species. Six species of frogs and toads, fifteen species of lizards and snakes, forty-one species of birds, and twenty-one species of mammals were judged of local ecological importance based on more than ten thousand observations.[7]

Because Henry was hired as an educator as well as a researcher, he had to coexist in an academic environment—and the biology department was not without its tensions. Edward Taylor, a tall, lantern-jawed herpetologist who'd been at K.U. for decades and taught comparative anatomy, was engaged in a bitter feud with Hall. Taylor was cordial to the new ecologist, but his habit of explosively saying things like "The boss is a crook!" embarrassed Henry, who felt that agreeing would be disloyal and was wary of the older man. A few years later William Duellman arrived with a Michigan Ph.D. as the new curator of herpetology, and would become among the twentieth century's most accomplished vertebrate biologists. His hard-charging personality didn't preclude a cordial working relationship with Henry, such that each named new species of lizards after the other.[8]

The 1950s and 1960s were years of intense transition in biology. Watson and Crick announced discovery of the structure of DNA in 1953, experimental approaches were increasingly emphasized in curricular and hiring decisions, and K.U. was a hotspot for ecology and evolution. Bright students came to Lawrence from all over the United States, exemplified by Paul Ehrlich, who was in Henry's first ecology course and later became a distinguished Stanford conservation biologist. Within a decade after Fitch's arrival his department included Charles Michener, who studied bee behavior; Robert Sokal and James Rohlf, pioneers in biostatistics; and Daniel Janzen, a charismatic leader in the emerging discipline of tropical biology—all of whom were eventually elected to the National Academy of Sciences. True to form, Henry steered clear of campus politics and only reluctantly attended departmental meetings; colleagues accurately regarded him as off in his own world.

For many years animal ecology was Henry's main teaching responsibility, but he also covered elementary biology and other classes. While he was on sabbatical in 1967–1968 the department reorganized and new faculty members were brought on board. Henry was switched to animal natural history, a less theoretical course, in which he adopted Grinnell's tactic of having undergrads read out loud from their field notes. Some of his graduate students worked on mammals, and there was even a chigger man, but most of them were herpetologists. Donald Clark, for example, conducted the first field study of a burrowing reptile, the western worm snake, while Michael Plummer's dissertation was about smooth soft-shelled turtles in the nearby Kaw River.[9] By all accounts his graduate mentoring style mirrored Grinnell's as well, with advice typically given in the form of questions and oblique commentary rather than detailed instructions, direct criticism, or unequivocal demands.

Dogged pursuit of natural history ran counter to an already widespread emphasis on experimental biology. Charles Elton and David Lack in England and Evelyn Hutchinson at Yale were posing sweeping questions about ecosystems, yet in 1960 Henry's longest monograph was titled *Autecology of the Copperhead*.[10] Others framed research in terms of synecology—the study of natural communities—or tested life history theory by comparing species, but although Henry summarized reptile population biology in several long papers, he didn't dwell on conceptual matters. His fieldwork emphasized the roles of particular species at the Reservation; perhaps jostling with the likes of Michener and Sokal was all the more daunting for a shy, unpretentious person who'd rather be looking for skinks. There were years, nonetheless, in which Fitch publications were more often cited than those of any of his K.U. colleagues.

Scientific productivity brought invitations to review others' work and contribute to symposia on vertebrate speciation and lizard ecology, as well as praise from peers. Philip Smith of the Illinois Natural History Survey wrote in *Copeia* that Henry's racer monograph was "a monument to effort and perseverance as well as productive original investigation and near-flawless literature research, the result of a fantastic amount of labor devoted to one species season after season for fourteen consecutive years. It consists, on the one hand, of imaginative probes into almost every facet of the snake's life cycle and its place in the environment and, on the other hand, of a methodical and machine-like precision in locating, sifting, digesting, and utilizing scattered bits of information from dozens of journals."[11] Forty years later, Jay Savage, himself a distinguished vertebrate biologist, would remark, "Fitch's long-term studies . . . on reptiles, are among the most meticulous and data-rich autecologies ever published."[12]

Social calamities couldn't be entirely avoided, even on the Reservation. Local farmers gossiped that Henry bred venomous snakes for release, a rumor given false credence by a grad student's death, probably from lightning. There was also the chronic threat of a run-in with Dr. Hall, as the Fitches always called him. The crusty museum director forbade nonnative animals from the Reserve—no horses, cattle, or pets—so Henry excluded exotic reptiles from his research and shot trespassing dogs that chased wildlife. Nonetheless, after their son Chester was born, Virginia successfully lobbied for a Siamese kitten. One day the Halls showed up, and after a few pleasantries, just as they were leaving, a loud *meow* came from behind the bathroom door. When Dr. Hall angrily brought up the no-pets agreement, Henry responded tersely that this animal was sealed off from the ecosystem. His boss issued an ultimatum: "Your job or the cat!" They gave it away but later had tame skunks, a great horned owl, a fox squirrel, and a pair of raccoons named Mabel and Jake.

Fitch family stories remind me of a band of happy gypsies. Henry was raised in a supportive environment, but he was always so alive in the moment, so unselfconsciously absorbed in nature, one could wonder how he managed as husband and father—after all, scientists aren't always exemplary in that regard, and our publications tend to be rife with apologies to loved ones. On the Reservation, though, with an office in the house and his lab next door, Henry was always nearby. He taught Alice to read, made toys, played games with the children, and told stories that have passed down to his grandkids. Virginia had him drop a set of stories about a naughty little girl lest they unduly influence their daughter, but the more enduring series, which featured "Old John" roaming the world for adventures with animals, began as Henry drove from California to Louisiana with little John beside him on the car seat.

Virginia coauthored papers with her husband, gathered observations of nesting summer tanagers through their bedroom window while pregnant,[13] and contributed in countless other ways during almost six decades of marriage. Family vacations included searching for gartersnakes in mosquito-infested Wyoming meadows when the kids were young, and the summer that Henry took a radioecology course at Oak Ridge they camped every weekend. Back on the Reservation he gave Chester fifty cents for slender glass lizards, out of which the boy reimbursed his friends who helped find the reptiles;

when I recalled that Grinnell paid a dollar for gophers, Henry cheerfully pointed out that the rodents were *prepared as specimens,* not simply caught and brought home. Lest there be any doubt about his gratitude, though, *A Kansas Snake Community,* Fitch's 1999 career capstone, was dedicated to Virginia and their children, "for all of whom natural history became a way of life."[14]

Like most of us who as kids read about reptiles, Henry yearned to see boa constrictors, green iguanas, and other charismatic tropical species. By the mid-1960s, when his children were old enough to travel, he was intrigued by a recently published, poorly supported claim that tropical animals reproduce year-round. Moreover, he had been doing catch-process-release at the Reservation for years; he'd never liked winter, when reptiles are inactive, and—unlike Joseph Grinnell, who famously limited his fieldwork to California—Henry was subject to wanderlust. As it happened, a consortium of U.S. and Latin American universities had just formed the Organization for Tropical Studies, and in 1965, at age fifty-five, he was the last professor enrolled in its now-legendary graduate ecology course. Daniel Janzen was on the faculty in Costa Rica that summer, and Henry watched with admiration for his colleague's verve as Dan said in a lecture that O.T.S. director Leslie Holdridge's habitat classification scheme didn't work and was fired on the spot.

There followed two decades of tropical biology, extending as far as Ecuador and conducted in the winters or when Henry was on sabbatical. Because tropical snakes are often nocturnal and he had poor night vision, research necessarily focused on other reptiles. The family spent a year comparing lizard demography at Costa Rican sites, and when Alice and her husband, Tony, were University of Oklahoma grad students they traveled for months with her father, surveying anoles in Mexico and Central America. His daughter was twenty when they first went to Costa Rica, and years later he chuckled at how young men volunteered assistance when she was along. David Hillis, then an undergrad at Baylor and now a University of Texas professor, accompanied him for a semester on another expedition.[15] And with grad student Robert Henderson, Henry studied iguana biology and conservation for both the right-wing Somoza and Marxist Sandinista governments of Nicaragua, as well as reptile ecology in Mexico and the Dominican Republic.[16]

Henry Fitch inspecting green
iguanas for sale as food in
an El Salvador market, 1979.
(Photo: D. M. Hillis)

Henry was a frugal creature of habit in the tropics, as he'd been at home—
his natural history was low tech, low cost, but high quality, and in a likely
unprecedented gesture he even returned some grant money after completing
a Caribbean project. Hillis calls the semester he skipped for their trip through
Mexico and Central America the most educational of his undergraduate
career; he was repeatedly impressed by the older man's field stamina, knowledge of local flora and fauna, and equanimity under duress. During four
months of living out of a pickup together, David saw Henry get mad only
once, when a border guard deliberately overcharged them. Henderson traveled
for six weeks with his advisor in Mexico, eating cereal with powdered milk
every morning and spam with boiled potatoes every night. Once when they
were out of milk and Bob suggested sandwiches, Henry replied incredulously,
"For *breakfast?*" and opted for water on his cereal.

As Henry moved through middle age his publications were increasingly
influential. Jay Savage's *Amphibians and Reptiles of Costa Rica* judged him "the

most important contributor ... during the 1970s" to that country's herpetology, one whose publications "form the starting point for subsequent research on lizard ecology."[17] In the 1980s, with Ph.D. student Fabián Jaksic, I combined his San Joaquin Experimental Range findings for broader analyses. Others had assumed, for example, that when many lizards have regenerated tails, the number of individuals actually eaten also must be high, but we demonstrated that predation by raptors, carnivores, and snakes wasn't correlated with injuries to specimens of the same five lizard species that Henry had deposited in M.V.Z.[18] Likewise, he'd noted that California ground squirrels, at risk of snakebite, were more skittish than those he'd hunted as a boy in Oregon, where rattlers were absent—observations that decades later led to research showing that squirrels enhance toxin resistance by teasing snakes into inoculating them with sublethal venom doses.[19]

In 1980 Richard Seigel, Fitch's last Ph.D. student, organized a symposium and subsequent volume to honor his "retirement." Among the presentations, Henry's mammalogist son John dedicated his to "an excellent scientist, friend, and father for the effects his enthusiasm, originality, and high standards have had upon family, students, and colleagues alike."[20] Over the next decade Henry finished a project on slender glass lizards, for which in thirty-five years he amassed 3,353 captures of 2,116 individuals (Ed Taylor, his irascible older colleague and a famously skillful collector, had found only a dozen).[21] He also published on rattlesnake roundups and began writing three books, including one on human populations called *The Last Doubling* and another, for children, about people who study snakes. The third, *A Kansas Snake Community,* begins, "My study of Kansas snakes involved the monitoring of local populations over fifty consecutive seasons from 1948 to 1997, with mark-and-recapture procedures."[22]

Professional colleagues bestowed more accolades after Henry retired. The university renamed the place where he had worked for decades the Fitch Natural History Reservation, agreeing that he and Virginia should stay there as long as health permitted. The Southwestern Association of Naturalists gave him its W. Frank Blair Eminent Naturalist Award. In 1998, the American Society of Ichthyologists and Herpetologists—which he'd joined as a grad student, remaining loyal despite "extraordinarily high costs and the unfortunate modern tendency to not hold meetings near campgrounds"—initiated the Henry S. Fitch Award for Excellence in Herpetology, and just as I was to announce the first winner, he bounded up to the stage and took the microphone. "Having as a young man published in *Copeia*," Henry told the audience, "this is my proudest moment as a member."

Physical stamina and emotional resilience are useful, even life-saving qualities for field biologists, and Henry always had them in abundance. He had run track and played tennis as a youth, winning the "under sixteen" tournament in Medford, and he still jogged into his eighties. Back in Oregon he'd assembled siblings and neighborhood kids for scrub soccer and baseball, and all his life he would relish ping-pong and basketball. On the Reservation, after long hours of fieldwork, fierce competitions took place behind the Fitch house on a dusty one-hoop court so bumpy that ordinary dribbling was impossible; on one occasion, Don Clark and I each jostled our professor so hard that he sustained a swollen, bloody eyelid. Thirty-five years later, when Henry, Alice, and Tony visited my Arizona rattlesnake study site, the older man traversed steep canyon slopes with the aid of a hefty wooden staff and finished the day with an impressive appetite for spicy Mexican food.

Two western rattlesnakes bit Henry during the San Joaquin work, as did several copperheads in Kansas, and he caught hepatitis in Ecuador; however, his only life-threatening field accident occurred barely two months past his eighty-ninth birthday. Out late on the Reservation, he became confused because of diabetes-related low blood sugar and in the gathering darkness careened off a twenty-foot-high embankment into a creek. A branch punctured one leg, and he endured the night partially submerged in frigid water. Meanwhile, sheriff's search dogs were hopelessly confused because Henry's scent trail was *everywhere*. He was shaken but conscious the next morning when rescuers helicoptered him from a nearby field to a Kansas City hospital, all of which he found embarrassingly overdone.

Virginia Fitch died at home after a brief illness, in November 2002, and the family scattered her ashes on the Reservation in a private memorial gathering. When Henry called me a few weeks later, he spoke of missing his wife tremendously and asked for assistance in conducting a telemetry project on timber rattlesnakes, to take his mind off her passing. He subsequently finished several more seasons of fieldwork on local rattlers and analyzed pellets to study the diet of long-eared owls.[23] Into his nineties he still gave talks to school groups, discussing plants, bird nests, and jars of live critters that he spread over an outdoor table. Sometimes he showed kids a big live snake too, just like when he was a five-year-old back in Oregon.

Henry accommodated to aging by enlisting collaborators, using an off-road cart, and reexamining decades' worth of data with new questions in mind. Alice and Tony helped with computer-produced illustrations for his publications. He watched nature programs and enthusiastically followed sports, especially Jayhawk basketball. In 2003 he underwent hip replacement, optimistic that fieldwork would be easier as a result, and within days gave his first PowerPoint presentation at a Kansas Herpetological Society meeting. The next spring Henry returned to Oregon with Alice for a new perspective on the family property and likely set an age record for going up in ultra-light aircraft. At that summer's A.S.I.H. meeting he spoke about timber rattlesnake research and checked in on fellow snake biologist Rick Shine, hospitalized from a cottonmouth bite on a local field trip. And some things never change. When predators killed our telemetered rattlers and I offered surveillance cameras, Henry simply put wire barriers and powder track stations around the remaining snakes. Low tech, low cost all the way.

Henry was fifty-two years old when I first met the Fitches, but he looked forty, soft-spoken yet energetic. A decade thereafter, as I began graduate school, the man was resolutely gathering field data, making his mark on tropical herpetology while younger ecologists were galvanized by blossoming theory. At a time when *American Naturalist*, "devoted to conceptual unification of the biological sciences" according to its masthead, was the hot journal in our mailboxes, he opted for more life history monographs. While others lumped snake stomach contents from throughout a species' range to test foraging theory, he demonstrated changes in copperhead diets at one site over a period of decades. Henry's path thus frames a riddle upon which most of us eventually reflect: Why does one do the work, and how does it matter? By the mid-1970s I was deep in my own Ph.D. research and keeping such questions well out of reach, but years later, while interviewing him for this book, I would discover that all the answers are intertwined.

FIVE

Medic

NATURAL HISTORY HAS ALWAYS HELD MY INTEREST even as other obligations suffered. During that year in the Georgetown funeral home I published the massasauga observations with George Oliver, two papers from research with Henry Fitch, and a fourth on Texas alligator lizard behavior with Ben Dial.[1] I studied Mediterranean geckos on the walls of a nearby tavern—once taking a date along in the hearse to collect them—and microscopy was among the few classes at Southwestern that I took seriously, because my favorite professor, Gordon Wolcott, let me examine those lizards' ovaries for a course project.[2] Along the way he scolded me about bad grades, but I was distracted by partying and after three years had an overall D+ average. In the summer of 1966 my downward spiral in academia was cemented by failing physics at the University of Texas, perhaps because I was enamored of a wealthy Odessa doctor's daughter—a match so improbable that right from the start our mothers predicted failure—traveling to see her on weekends rather than study.

As I emerged from class at noon that August 1, I heard popping sounds that I assumed were firecrackers at an antiwar demonstration. Then I learned on my car radio that someone was shooting from the clock tower. A deranged Charles Whitman slaughtered sixteen people before two cops killed him; that evening some of the victims were laid out in an Austin mortuary when I went back to pick up a Georgetown resident who'd died of natural causes. By then I'd already seen dozens of strangers crushed by the loss of loved ones, and several people close to me had died as well. Grandpa Gibson succumbed to cancer when I was seventeen, leaving me inconsolable, and soon thereafter, just as we moved to Fort Worth, my Missouri herpetologist friend Paul Anderson suffered a fatal heart attack. Later, when a wreck killed our Warrensburg neighbor

Texas alligator lizard, Chisos Mountains, Brewster County, Texas. This species inhabits the Hill Country and Trans-Pecos parts of the state. (Photo: H. W. Greene)

Stan Fisher and two of his kids, I'd begged my mom for explanations. How would Mildred survive the loss of her husband and children? Why is life so unfair? As the summer ended, with the tower sniper's carnage fresh in my mind, that last question felt more and more unanswerable.

In short order the debutante dumped me and my old Volvo conked out, so in desperation I transferred to Texas Wesleyan College, where Daddy taught. I'd hit bottom academically and financially. I first sold clothes in a department store, then drove a laundry truck, but both jobs were boring and paid poorly. Seeking cash and something edgy to break the tedium of school, I took an advanced first aid course and hired on with an ambulance company. Our manager tossed expired donor blood on a sheet to accustom us to the startling mess we would encounter, and weekly training sessions included a navy film on emergency childbirth. I worked round the clock four days a week, attended classes the other three, and made excellent grades. Emotionally, though, I was well on the way to shutdown, and my memories are staccato vignettes, mostly devoid of feelings because I was largely unaware of having any. We handled mangled, charred, and decomposed bodies, but I got physically ill only twice, when the bowels of dead people emptied on me.

Gold Cross worked two-person crews out of a central headquarters and four suburban stations. We started at 7 A.M., taking a few minutes to inventory first aid supplies, refill oxygen tanks, and check out our vehicle. Entire shifts sometimes passed without an emergency, but on weekends a crew might pull several traffic accidents, a shooting or stabbing, and a heart attack. My assistant was often a stout, good-natured T.W.C. student from Central Islip, New York, nicknamed Slim, who kept me laughing with wisecracks delivered in a heavy Long Island accent. Between calls we studied, watched TV, and slept. Sooner or later a buzzer would go off, followed by the loudspeaker: "Three-car accident, northbound Loop 820, south of Highway 121." We'd run out lighting cigarettes, then I'd flick toggles for the emergency beacons and turn on the siren as we rolled down the driveway. Within minutes we'd be applying compression bandages, putting on inflatable splints, and inserting "resuscitubes" over the tongues of unconscious patients.

Slim and I worked in a chaos of flashing lights, clanking tow-truck chains, and moaning victims; we treated patients amid the worrisome and vaguely nauseating smell of gasoline, while all around us blood was pooled on floor mats, smeared on windshields like scarlet finger-painting. Strangers were literally and figuratively in our hands, with rumpled clothes, smashed glasses, arms and legs limply askew. One woman kneeled over her wounded boyfriend and kept us at bay with a knife until police arrived, but usually injured people were cooperative, even poignantly trusting. Sometimes their eyes darted about as if an invisible, dangerous *something* were out there; other times they seemed oddly detached, as if contemplating some distant memory. We'd unload at a hospital and fill out paperwork, then return to the station, replenish our first aid supplies and grill steaks for dinner.

Patients usually required only straightforward stabilizing procedures, but occasionally complications arose. Once I had a compress on a man's head gash when bystanders alluded to another wreck victim in a nearby ditch. Because my guy's bleeding was controlled and another ambulance would've taken twenty minutes to reach us, I sent Slim into the darkness. Suddenly someone jerked my hand off the dressing, shouting, "I'm a nurse and this man's injured. You'd better get him to a doctor!" I muttered, "Oh shit!" and pushed her aside, reapplying the bandage and reassuring the victim. Meanwhile Slim concluded that no one else was hurt, and by the time we reached a hospital the nurse had phoned in about my behavior. Another night we picked up a tiny woman who'd sought refuge from her husband's blows behind a refrigerator. She weighed less than a hundred pounds but was so wrought-up from physical and psychological battering that two big cops had to help us gently wrestle her onto a gurney.

We were legally bound to treat all intact patients even if they weren't savable. I gave CPR to numerous heart attack victims with open airways and good skin color, for example, but none of them revived. One night a hammer-wielding thug attempted to assault an elderly lady in a seedy apartment building. She, however, fired a .38 point-blank and left him with streaming neck wounds and no pulse. The stairwell was too narrow for a gurney, so Slim pulled him to a sitting position, I grasped his wrists from behind and under the armpits, and with Slim hefting the legs and a lifeless head flopping against my chest we crab-walked him down four long flights of steps. He was dead on arrival, and I had to change out of my blood-soaked jumpsuit before the next call. Externally less impressive injuries also could prove fatal, as in the case of a suburban couple, she sobbing and the husband stoic with a blue-edged .22 caliber hole in his chest. He spoke calmly to police—"Well, we were arguing, and the wife, uh, accidentally shot me"—then got on the gurney, lapsed into unconsciousness, and died just as we unloaded him at the hospital.

One Friday night we were called to an embankment on the city's outskirts and found a policeman bent over a car hood giving mouth-to-mouth to a toddler. The two-year-old had turned cyanotic from a penicillin shot, and during a frantic drive for help her father lost control and rolled the vehicle. I took over from the cop, tilted the girl's chin back, and checked with a finger that she hadn't swallowed her tongue; then I sealed my lips over her nose and mouth and began soft, puffing exhalations. I could feel the tiny chest expand and subside—*that's it, come on darlin'*—and every few breaths I pushed on her sternum with two fingers, compressing the heart and causing blood to flow. The cop's flashlight beam revealed color in the child's face as I continued CPR and struggled upslope with her. Slim took over as I proceeded to guide the hysterical mother into the front seat and drive to the hospital; within minutes, after handing her daughter over to a physician, we collapsed outside the ER. A nurse soon emerged with the news that we hadn't saved the little girl, and even now I can remember her blond curls and pellucid skin, the sour milk taste of her mouth. I still recall the abysmal grief on that young woman's face and how cold the green corridor tiles felt when we slumped against them weeping.

All that weekend I thought about quitting, but the next Tuesday, on a sunny April afternoon, my crew helped deliver a baby. Father and grandfather were pacing curbside in front of a ramshackle house, exclaiming, "She's having it, she's having it!" An eighteen-year-old woman lay on a couch in the front room, prepped that morning by an intern who'd decided birth wasn't imminent and sent her home. Now she grimaced and groaned, her puffy face flushed and beaded in sweat. I parted her legs and encountered multicolored swirls of liquid, a taut three-inch vaginal opening, and the wispy hair of a crowning infant. *This baby is coming now!* We helped the woman rise up against some pillows and laid out clean towels, all the while talking calmly to each other and the bystanders. Then as she screamed and bore down, I held my palms upward to cradle the head and pressed fingertips against her perineum, to slow birth and prevent tearing.

The baby came sliding out fast once its head emerged—*with the umbilical cord wrapped around his neck!* I grasped the slimy, vein-swollen noose and pulled enough slack from the birth canal to untangle it. *No problem!* Next we suctioned fluids out of his mouth and nose with a bulb syringe, dried him off, and wrapped the wrinkled, blotchy newborn in a small blanket. By then the cord had collapsed, so I applied a two-part "umbili-clamp" that severed it, leaving a short piece attached to the infant and a longer one draped upon the woman's belly. Soon everyone was smiling and the whole family was crying as we wheeled new mother and healthy, nursing son up to the same doctor she'd seen earlier that day. I was grinning too, but back out in the hospital parking lot I had to lean against the ambulance doors and steady my hands to light a cigarette.

That next fall, during my last semester at T.W.C., the mentally ill teenage son of my biology advisor back in Georgetown shot his sleeping parents and sister. Dr. Wolcott's murder evoked surprisingly little emotion in me, perhaps because I was already benumbed to tragedy and otherwise distracted: I was dating Donna, a woman I'd met at Southwestern. We married in early 1968, within weeks of graduation. The Vietnam War was in full swing, and when my draft notice arrived I enlisted for an extra year in return for the army's guarantee that I'd be a medic.

Basic training at Fort Bliss lasted two months in the El Paso summer, during which I lost twenty-five pounds from the Spartan diet and intense daily regime. We did calisthenics, ran miles under the weight of a steel pot helmet and light pack, and took classes. After breakfast a captain would lecture on the Uniform Code of Military Justice or the Geneva Convention rules for treatment of prisoners. Then a drill instructor showed us how to dispatch the enemy with rifles, bayonets, and bare hands, all the while shouting, "Kill! Kill! Kill!" Day after day, week after week we gulped down meager rations, choked through tear gas, hurled grenades at piles of tires, and low-crawled under concertina wire, around exploding bunkers, and beneath machine gun tracer fire.

A recruit's life made Boy Scout hikes seem trivially easy, but there was a dichotomy in our ranks: draftees just out of high school were in good physical shape but rebelled at mental hazing, whereas us flabbier "college boys" may have found exercise grueling, but I just laughed inside when a "D.I." leaned his hat brim against my forehead and shouted obscenities. Training was couched in terms of "when you get to 'Nam" and "if the Viet Cong overtake your foxhole," and of course widespread antiwar sentiment was on our minds, however rarely mentioned. One June morning we stood at attention as a sergeant laid out the day's schedule, then added, "Some of you might like to know Robert Kennedy was assassinated last night. Personally, I'm glad the son of a bitch is dead."[3] Official justification for the conflict kept changing, and I had deep misgivings about it myself, the more so for already having handled lots of gunshot victims, but by the end of basic I thought I could kill if it came to down to either that or someone killing me.

Many of the men in my company at Fort Bliss went on to advanced infantry training and a tour of duty in Southeast Asia, whereas I was sent to San Antonio for medic school. Our first class consisted of a film about army medicine in Vietnam, which followed several soldiers from injury to treatment and rehabilitation. In one scene a mortar blast reduced a sergeant's leg to a smoldering stump; under withering small-arms fire, a medic low-crawled to the injured man, applied a tourniquet, and pulled him to cover. Another case depicted a surgeon pulling an intact slug out of a soldier's eye, at which point one of my fellow trainees fled the room gagging. Over the next few months we made hospital beds, emptied bedpans, took blood pressure, and gave injections; then I learned to chart and clean teeth, take X-rays, and make gelatin impressions for fitting dentures.

After I'd completed the medic and dental assistant courses, my first assignment was back at Beaumont General Hospital in El Paso, where our patients had disfigured faces, missing limbs, and other dramatic war injuries. Given the likelihood that I was headed for Vietnam, distractions were welcome. Checkered whip-tailed lizards foraged on the hospital grounds, and an officer showed me Komodo monitor photographs from his travels in Indonesia. Off duty I read scientific journals, worked on manuscripts, and kept in touch with friends like Jay Cole, who had moved to Arizona for a Ph.D. Weekends I prowled the Chihuahuan Desert, where I caught a New Mexico whip-tailed lizard for Roger Conant to illustrate in his new *Field Guide* and a crevice spiny lizard that later gave birth in my terrarium. I found my first black-tailed rattlesnake and marveled over its beadwork appearance, the consequence of each scale having only a single color.

Then I got the sort of lucky break that haunts one's later years. Nine medics in our clinic received overseas orders within a week, seven to Vietnam and two, including me, to Germany. Up to that point my job at Beaumont had been dentistry, enlivened by occasional duty as a surgical technician, but one summer day, as my transfer overseas loomed, I noticed a putrid, vaguely sweet odor and remembered the morgue downstairs. Just then an officer who knew of my funeral home experience sought help identifying a couple of bodies, so off we went, wearing flimsy paper masks against the stench. Leaking gas had killed a private and his wife in a house trailer; they'd lain there more than a week and were so badly decomposed that we could only distinguish them by her long hair. I used wooden tongue depressors to force open the melted holes that had been mouths, then stared at a ceiling light while the captain compared their teeth with dental records. That was the last time I winced at rotted human flesh, but the fate of those other seven soldiers still clouds my memories of military service.

For almost two years Donna and I lived in Frankfurt am Main, birthplace of Johann Wolfgang von Goethe, where I was assigned to the medical detachment of a tank division. The half day a month that soldiers were allowed off I visited the Senckenberg Museum of Natural History, itself founded by Germany's most famous poet. On weekends I discovered fire salamanders under logs in the Taunus Mountains, snatched yellow-bellied toads from roadside pools, and photographed an adder basking in a Bavarian bog. Vacations were devoted to sunnier, more biologically diverse Mediterranean landscapes. We camped in an olive grove along the Arno River in Florence, rode motorbikes through the hills above Nice, and in Madrid's Prado Museum I was taken aback by Francisco Goya's darkly powerful *Tres de Mayo,* reminded by his faceless Napoleonic executioners of my own run-ins with human cruelty.

While we were on vacation in Spain, I lifted a flat rock and was greeted by the open-mouth threat of a sixteen-inch-long ocellated lizard, and under smaller stones I found Iberian worm-lizards, recognizable by their vestigial eyes and rings of rectangular scales—the first individuals I'd encountered of the elongate, limbless Amphisbaenia. Back in Frankfurt I bought a South American checkered worm-lizard in a pet store and noticed that when I touched its head or body, the burrowing reptile waved its black-and-white blotched tail. On my next afternoon off I reviewed literature in the Senckenberg and realized that other amphisbaenians and snakes with similar displays fell into two groups, implying different ways of thwarting enemies. Some species seemed to use bright colors to advertise noxious qualities, while others distracted predatory attacks to a specialized tail, away from the more vulnerable head and neck.

During trips to natural history museums in Leiden and London I tested those ideas by examining hundreds of preserved specimens of two Asian species. Red-tailed pipesnakes look like deadly cobra relatives yet rarely had scars or incomplete tails, consistent with a bluffing function for their gaudy displays, whereas more than 50 percent of brown sand boas had stout, drab tails that were punctured, scarred, or incomplete. Published observations of one with a bleeding tail, otherwise unscathed and surrounded by jackal tracks, also agreed with my hypothesis that that species decoys injuries to its unusually bony, expendable hind end. Checkered worm-lizards apparently use a mixed strategy: their thrashing black-and-white posteriors mimic a venomous coralsnake and, should that fail to deter, direct insistent enemies to a disposable tail.[4]

We returned to the United States in May 1971 and spent my first night as a civilian with Roger and Isabelle Conant in the New Jersey Pine Barrens. Next day we headed for the Fort Worth area, where I'd been accepted for a master's program at the University of Texas at Arlington. That fall I began writing up the tail display project, studying ecology and genetics, and carrying out thesis research. My world felt intact despite the deaths of Grandpa Gibson and Paul Anderson, the loss of our Missouri neighbors and Dr. Wolcott, and six years as a medic—after all, I'd missed Vietnam and, miserable grades notwithstanding, I was in graduate school. Then one Saturday morning in September 1972, I came home humming and juggling groceries, propped open the back door with a knee, and looked for my wife's familiar face.

"Your mom's on the phone," she said in an unfamiliar tone. *Maybe they're inviting us for dinner—but what's that anticipatory firmness in Donna's voice?*

"Hi, Mother, what's up?"

"Harry, Marsha was murdered Wednesday night." *Those green eyes, that impish smile—we'd been just kids, so naively intimate, never really broke up, and each married someone else...*

"Today's paper says police found their car in an intersection . . . " *I've seen those dark holes in a woman's temple, the bloody matted hair.*

"Marsha was in the front passenger seat."

For years the scene was all so clear and satisfying in my mind's eye: Marsha's husband had killed her, but he put on a show of grief and the cops, running out of leads, didn't have enough evidence to arrest him. Months later, a winter storm gathers as he stands hatless in a Texas panhandle farmyard, bent over tree limbs with a chainsaw, and on this desolate afternoon the man is cutting firewood as if nothing had happened, as if he hadn't shot the mother of their infant son. His single bit ax is propped against a stump and branches are piled nearby, trimmed off for kindling before he starts cutting bigger pieces into stove logs. There is nothing to hide my approach—not another tree in sight, and the outbuildings are not close enough for this purpose—but Lyndy never hears anything above the roaring saw motor, never sees my upraised arms, never knows what hit him.

That morning after my mom called I hung up and struggled with heartbreak. Having put away the groceries, I remembered a couple with whom Marsha and I had been friends. I steeled myself for a few minutes before dialing their number in a nearby town. It was a relief to hear Robin answer instead of his wife, who'd been Marsha's roommate at Southwestern, but Dee came on their other phone, and as I stammered out details of the murder she began shrieking, "No, oh please no!" Two days later I sat in stunned, furious disbelief as a minister stood over Marsha's casket and proclaimed this as God's plan. At the graveside, mourners stood in line to speak with the family. I approached her mother and brother, who didn't recognize me at first, with my longish hair and droopy moustache. Then she raised both hands, as if to cover her mouth, and said, "Look, it's Harry . . . "

I swing from high and to the right, so hard that later my shoulders ache, and the ax lands with a loud thwack, stuck so deep in Lyndy's head it yanks me along when he collapses. I scream out, "You sorry ass bastard!" He falls over, twisting and scissor-kicking like a pithed frog. Then his eyes start jerking around, so I use a boot to hold his face into the snow and look away, listening to the wind pick up, suddenly aware of labored breathing, the biting cold on my wet cheeks. After a couple of minutes he stops twitching. Blood is fast soaking the snow around his head, and a wet stain spreads high on the front of his pants. "Lyndy," I howl, "goddamn you motherfucking son of a bitch!" I'm wearing gloves and leave no fingerprints, meet no one on the road back to town, and by nightfall the snowstorm obliterates my tire tracks. Of course I never mention a word of this to anyone nor suffer a moment's guilt.

Life tumbled on. Donna had become a born-again Christian before join-ing me in Germany, and as our religious beliefs diverged I became more emotionally removed. We kept up a cheery facade for the next few years but divorced when I finished my master's and moved to the University of Tennessee.

For more than a decade I had no contact with Marsha's family and scarcely spoke of her, even to close friends. In 1985, though, as an organ played *Finlandia* and I walked up a church aisle during my own father's funeral, her dad reached out from a pew, took my hands, and began sobbing. A few years later Mom and I ran into Robin and Dee at a Dallas cafeteria, and after a round of introductions Dee said that once in a while she still has a good cry for Marsha. Then as we left my mother reminded me that police had found Lyndy slumped over the steering wheel, mortally wounded and clutching the .22 pistol he'd used to kill his estranged wife. Minutes earlier their child had been left with friends; a fifteen-page note on the seat read as if the couple had acted jointly, but a judge emphatically ruled the incident a murder-suicide.

"Well, honey," my mom said, unaware of the anger I harbored, "maybe it's just as well Lyndy took his own life."

There followed Ph.D. work and an academic career, as I became absorbed in the practice of natural history. Along the way, I've taught at three fine uni-versities, sagged under academic fads and skirmishes, then bounced back by refocusing on organisms, having learned to trust Alfred Russel Wallace's emphasis on life's paths instead of its peaks.[5] I studied rattlesnakes with Henry Fitch as both he and the Museum of Vertebrate Zoology approached their hundredth birthdays. I've been counseled in tropical ravines and high Andean passes, mesmerized by the sexy elegance of maned wolves on the Brazilian *cerrados,* and humbled by a pissed-off African elephant. Attention to nature's intricacies has convinced me that we are no more permanent than earthworms or oaks, and rather than hope for eternity I revel in the moment and savor relationships, seeking joy in work itself. Revenge and sorrow no longer weigh so heavily either, and when memories well up I celebrate the special qualities of lost loved ones. I recall a young woman's cheerful smile, reflect on my father's decency and admire how brother Will resembles him.

Still, though, as Jim Harrison said in *Legends of the Fall:* "Who reasons death anymore than they can weigh the earth or the heart of beauty?"[6] On a brisk fall evening I open a fine Chilean red and crank up the stereo with vintage rock and roll. Familiar guitar riffs soar, the fireplace crackles, and an hour goes by. I maintain an absentminded stare while dancing flames scour the darkness, as if they might vanquish some elusive menace lurking in the shadows; shifting about on the couch, I imagine lions roaring outside a Pleistocene cave. Another hour passes. By now the wine has warmed me inside, loosened my practiced grip, and Jackson Browne's "Barricades of Heaven" draws out my reluctant emotions. Later still, when embers have waned and composure returns, Riley puts his big yellow dog head in my lap and solicits a hug that I need far more than he does.

There is an enormous maple tree outside our house in upstate New York, and tomorrow morning, against a chilly sky, its shimmering orange leaves will be incandescent in the predawn light. Soon enough I'll slide into silence with countless others who have already come and gone, but for now we're riding high in this colorful earthly riot.

PART TWO

———

Conversing with Serpents

SIX

Graduate School

ACADEMIA ISN'T FOR EVERYONE and there are many ways to be a naturalist, so I ask career-seeking undergraduates two questions: Which activities do you enjoy, and what kinds of accomplishments would give your life meaning? Jane Goodall, among the most publicly admired of all biologists, provides a familiar example for distinguishing between *what one does* and *how it matters* in broader contexts—because if the discomforts and isolation of her early fieldwork had been intolerable, the famous chimp watcher's scientific stature wouldn't have followed, and her later global activism would not have been possible. By the same token, for some folks, unlocking secrets of coral reef fishes or rainforest birds just isn't fulfilling, no matter how good they are at doing it. Clearly, the trick is finding short- *and* long-term satisfaction, be it as herpetologist or humanitarian, police officer or stay-at-home parent.

Next, I describe to ambitious students what lies ahead. Aspiring professors must gain admission to graduate school, choose an advisor, and obtain financial support; ten years may be spent completing the Ph.D., working as a postdoctoral fellow, and landing a job. I jump-start the process by asking if my advisees prefer certain organisms ("goldfinches" and "no" are among many possible answers) and particular disciplines (physiology isn't for the squeamish). Then they tally authors of inspiring articles in publications like *Journal of Herpetology, American Naturalist,* and *Conservation Biology,* so I can offer feedback on potential mentors. I also encourage taking time off before advanced studies and emphasize that university jobs aren't necessarily more satisfying or influential than others. Students can see what works for me, but I want them to consider whether they might be happier and do just as much good in the world as an environmental consultant or a grandmother.

As it happens, no such careful planning dictated my own route. During the high school internship with Henry Fitch I'd learned that a Ph.D. could lead to studying reptiles, and three years in the army left me hungry for more education, but I had no core interests with which to choose among specific options. Worse still, my undergraduate record was utterly unacceptable for established doctoral programs. Given those highly problematic circumstances, I was lucky to find two supervisors at a time when most biologists thought snakes unworthy of serious study, and all the more fortunate that they allowed me to go my own way.

This is the first of five chapters focusing on my favorite places and the creatures that inhabit them. It's also about studying and teaching natural history, as well as finding one's path more generally. Academics often look back fondly on graduate school, but I'll wager most of us had little inkling of how intellectually and emotionally transformative it would be when we sat down in that first seminar. Watching frogs and snakes, conducting experiments and analyzing data, arguing about the conceptual foundations of biology and writing papers—these activities pervaded our waking hours, even our dreams, but with luck we also learned that happiness is about more than science. Only much later would I reflect on the ways in which mentors and friends influenced so many aspects of those years.

William Pyburn, my master's thesis advisor, bridged a penchant for observation with the emphasis on hypothesis testing that permeates modern science. Bill exhibited remarkable artistic abilities as a child and by the age of eight was taking art classes at a Houston museum. As an inquisitive East Texas teenager he watched broad-headed skinks and copperheads among the loblolly pines and willow oaks; he prowled cypress-lined stream banks in search of snappers, stinkpots, and other turtles, sitting spellbound as a swamp rabbit hopped by and then swam away. Once Bill surprised a basking alligator, and decades later he remembered how the reptile obscured its escape by lashing up a muddy cloud with its tail. A born collector, he even assembled an entire horse skeleton on the family lawn from bones found in his wanderings.[1]

Bill went on to college and a Ph.D. after serving as a navy medic in World War II. He was married to Wanda, his high school sweetheart, for fifty-seven years. Frank Blair, his University of Texas professor, had trained at Michigan, inventoried the faunas of several West Texas mountain ranges after moving to Austin, and then investigated population dynamics and speciation in toads. Under Blair's tutelage Bill studied mate choice by spiny lizards for his thesis, then staked out a dissertation on evolutionary genetics of color pattern in cricket frogs.[2] He'd always been thin and wiry; when another graduate student drew up an identification "key" for members of their group, Bill was described as "disappears when turned sidewise, runs around in the shower to get wet." Intellectually rigorous and surprisingly iron-willed, he could talk with equal facility about Bertrand Russell's philosophy, the history of jazz, and amphibian evolution.

Bill was always an artist at heart. His drawings and paintings were marked by attention to details, a propensity also evident in his scientific endeavors. He wrote in neat longhand, preparing an outline and then formulating sentences, even entire paragraphs, before putting pen to paper. His drafts required little or no revision. He was a dedicated rationalist and problem solver, such that in his waning years, when Parkinson's disease prevented making the exact lines needed for bird portraits, he switched to abstracts. Bill suffered from chronic depression, accentuated by the confines of campus and city, but he reverted to a more carefree demeanor in the field. Attentive to small flowers and other nuances of a desert canyon, he laughed easily and his eyes twinkled; preoccupied with frogs and birds in an East Texas pine woods, he walked like a man on good terms with the whole neighborhood.

At Southwestern I'd read Bill's publications about Mexican treefrog behavior,[3] then noticed his Arlington State Teachers College address and made an appointment to meet with him the next time I came home to Fort Worth. He shared a small office with another professor, Thomas Kennerly. At my knock the door crept open, revealing a Darwin-like figure with a largish bald head, dark bushy eyebrows, and a cropped mustache that accentuated his somewhat pinched mouth. Bill wore a bow tie and smelled like pipe smoke. "Are you really interested in biology," he asked, rocking on his heels and sticking out one hand with an air of skeptical detachment, "or just a snake collector?" I wasn't quite sure what that meant, but after some pleasantries he ushered me down a hallway and opened a closet door labeled Collection of Vertebrates. From floor to ceiling the shelves were chock-full of amphibians and reptiles in jars of alcohol, and in our subsequent meetings it seemed as if each specimen posed a question.

I visited Bill often after transferring to Texas Wesleyan College in the fall of 1966, encouraged that this shy, bright man thought me worth his time. One day he pulled out a pickled snake that in life was orange with a black head, black neck ring, and black spot at the base of the tail. Then, fixing me with his "What do you think?" look, he explained that the specimen came from the Sierra de Los Tuxtlas in southern Veracruz, Mexico. "Probably a female, based on the slender tail," I responded, hoping to impress him, "but what is it?" He handed me another jar containing two collapsed eggshells and two hatchling snakes, one patterned like its mother and the other with black bands all along its body. "Those two oddly colored animals won't key out to any known genus," Bill said with a quizzical grin, "but the banded hatchling is some sort of *Pliocercus*."

After a few hours of library research, I'd learned that numerous species and geographical races of *Pliocercus* had been named based solely on their colorful markings, making it all the more intriguing that those Tuxtla snakes differed dramatically within a single clutch of eggs. What if, in fact, only one species occurred in all of Mexico and Central America? Bill generously suggested that I examine more museum specimens, then publish the results during my upcoming military hitch.[4] While I was overseas in the army, he encouraged me to return for a master's degree at what had become the University of Texas at Arlington, and when my grades warranted only probationary admission he rescued my sputtering career by arranging a teaching assistantship.

That fall of 1971 Bill let me dangle in uncertainty over a thesis topic, as had Joseph Grinnell with his student Fitch, and coincidentally I almost emulated Henry's solution. Most of whatever biology I'd previously learned had faded, and although my research soon focused on behavior, that first semester I scarcely knew what it encompassed. I turned to Texas alligator lizards for lack of a better plan, having earlier discovered their parental care with my friend Ben Dial,[5] but soon a boldly colored book and some even more garishly marked snakes diverted me. Thirty-five years would pass before I published anything else on my teenage favorites.

The orange cover of Irenäus Eibl-Eibesfeldt's *Ethology* drew me into an account of how Niko Tinbergen, Karl von Frisch, and Konrad Lorenz studied what animals *do* from a naturalistic rather than a human-centered, psychological perspective.[6] Beginning in the 1930s they focused on insects and birds to address four questions, formalized by Tinbergen in 1963:[7] How is behavior controlled by motivational and sensory mechanisms, such as hunger and taste? How is it shaped by genetics and development, including experience? What roles does it play in survival and reproduction, and thus how does it relate to individual ecology? Finally, how did it evolve over geological time? Although the ethologists' vision was explicitly comprehensive, they emphasized instinctive behaviors—those that first appear "full-blown," with no learning involved—and they stressed assembling a detailed behavioral inventory, called an ethogram, prior to conducting experiments. Lorenz himself had trained in comparative anatomy, and early on he studied how behavior originated and changed during the evolutionary diversification of ducks and geese.

Tinbergen's digger wasps, von Frisch's dancing bees, and Lorenz's water fowl were on my mind while I wrote up the European museum project on snake tail injuries and delved more deeply into ethology. Baylor University's Frederick Gehlbach had just published on serpentine defensive displays,[8] so one day I made the two-hour drive to Waco to chat about our mutual interests. As the visit ended Fred offered me a live Texas coralsnake, and having never kept cobra relatives, I welcomed a chance to observe the thirty-inch-long gem. Back at U.T.A. my captive crawled under moss in its terrarium, resurfaced, and then attentively followed the invisible path of a rough earth-snake I'd dropped in as food. The earthsnake succumbed within minutes of being bitten, whereupon the larger serpent, after deliberately walking its jaws along the prey, consumed it headfirst.

That evening I read everything available on coralsnakes and prey trailing. Knowledge of snake diets had mostly come from chance observations and by forcing animals to regurgitate, but because snakes eat infrequently and their stomachs are usually empty, those methods require the sort of long-term studies pioneered by Fitch. Lugging journals to my library carrel, I sifted through a century of *Zoological Record* and lingered over Karl Schmidt's 1932 paper in *Copeia* on the gut contents of coralsnakes.[9] Schmidt even identified a new species of burrowing serpent among their prey, but evidently no one had followed his lead and really tapped into museum specimens—accumulated by many collectors over the course of decades—to conduct a detailed study of a single snake species' diet over its entire range.[10]

Around midnight I charged back to the biology building and down into the ever-growing Collection of Vertebrates, where I pulled out several jars of pickled coralsnakes. Then I carefully slit each one's belly, as I'd done as a teenager with the skinks at K.U. and massasaugas in Fort Worth. Although the first few coralsnake stomachs were empty, within the next couple of dozen were several food items. Prey were often partly digested, but by comparing remnants with intact specimens of all potential species, I identified them, determined sex, and estimated length and mass. Clearly I was on to something. Preserved coralsnakes without prey required only a few minutes to process, but those containing their last meals took half an hour, and I barely finished in time for my own breakfast. That same morning I asked Bill about changing my thesis topic to coralsnake feeding, and with his enthusiastic approval I set about visiting other collections and accumulating more live animals.

Fred Gehlbach also had demonstrated that blindsnakes follow ant trails,[11] and he encouraged me to modify his research setup so as to examine coralsnake hunting tactics: a few field observations suggested that they actively forage, and I was to test the hypothesis that they used chemical cues to find food. Prior to each experiment I encouraged a prey snake to crawl around an octagonal alley, its movements limited by cardboard walls on a cloth substrate that was washed between trials. As predicted, a hungry coralsnake, released in the yard-wide arena after the alley and smaller snake had been removed, inevitably followed the invisible spoor of its prey. Trailing behavior was especially obvious at the turns, which coralsnakes overshot by a few inches, then relocated by side-to-side head movements and tongue-flicking. After a few minutes an undulating coralsnake looked like lacquer-banded sine waves moving around an octagon.[12]

My captive coralsnakes seized other snakes willy-nilly but always swallowed them headfirst, as had been the case with 96 percent of 150 prey items in preserved specimens. The advantages seemed straightforward—a blunt head was easier to grasp than a slender tail, jaws more easily passed over backward-projecting scales and body parts—but how, I wondered, do beady-eyed coralsnakes distinguish the head and tail of their prey, hidden a foot or more away in leaf litter? Captives ingested short pieces of prekilled snakes from the front, proving that neither head nor tail was necessary and that a directional cue existed all along the prey's body. Next I removed skins of dead snakes, in the manner of peeling off a stocking, and replaced them on the carcasses backwards. Coralsnakes moved to the "front" of those experimental items—actually the prey's tail—supporting my hypothesis that they used backward-overlapping belly scales to find a prey snake's head.[13]

Fitch had whetted my wanderlust with publications from his Latin American sojourns, and Bill enthralled me with tales of driving a pickup from Texas to Colombia—his wife, Wanda, teenage daughter, Karen, and three students cramped in among supplies. The Pyburns spoke of exotic birds, strange night sounds, and rainforest trees so large they camped among the wall-like trunk buttresses. Ever the careful observer, Bill wrote in one letter from the field of a tiny green kingfisher flying low over a puddle, catching a single tadpole on each pass. And once, attracted to a commotion in the canopy, he watched an agitated woolly monkey snatch something off a branch and hurl it downward, stared as what looked like monkey shit materialized into a blob with outstretched limbs, then caught what proved to be a much sought-after canopy treefrog!

My own first tropical adventures were modest—a driving marathon to Guatemala with new friend Jonathan Campbell from the Fort Worth Zoo, a trip to Veracruz with two other Pyburn students—and exemplified the mystical allure that draws naturalists toward the Equator. Now I remember tense border crossings and vividly garbed Indian women, crested basilisk lizards and mosslike treefrogs, and one prophetic incident. On my knees in wet pasture, tearing open rotten logs with a potato rake, I spied the thrashing orange coils and black spots of a Tuxtla coralsnake. Having seen similarly "abnormally colored" specimens in museums, I used tongs on the snappy, eighteen-inch-long serpent, whereas my companions, unaware that not all coralsnakes are banded, would have simply grabbed it, with potentially lethal consequences. For me that animal sparked a new research interest.

In 1867 Alfred Russel Wallace, building on his fellow explorer Henry Bates's discovery of noxious "models" and palatable but protected "mimics" among butterflies, suggested that the brightly ringed colors of venomous coralsnakes warn predators and that harmless serpents gain protection by looking like those deadly species.[14] I'd read Wolfgang Wickler's *Mimicry in Plants and Animals,* however, and was impressed by his argument that coralsnakes couldn't be models because fatally bitten predators wouldn't survive to later avoid nonvenomous mimics.[15] By then Jay Cole, my friend from K.U., was at the American Museum of Natural History, and when I voiced skepticism about mimicry while examining coralsnake stomach contents there, he asked, "What if there doesn't have to be any learning? Maybe they can be models even if their enemies die."

I'd studied those strange spotted *Pliocercus* while in the army and seen the likewise patterned Tuxtla coralsnakes, so Bill and I collaborated on a theoretical paper about innate recognition of venomous serpents. We imagined a population of predators that genetically vary in tendencies to attack red and black patterns, with those more likely to avoid coralsnakes being favored, while their offspring, with no previous experience, avoid harmless snakes resembling the deadly models. Neither of us had ever had a manuscript rejected, but *American Naturalist* and *Evolution* turned down that one for lack of supporting data. Our essay was finally published in *The Biologist*,[16] and later studies confirmed Jay's idea that predators innately avoid coralsnake patterns, as well as the central tenets of Wallace's coralsnake mimicry hypothesis.[17]

My pleasure proctoring biology labs was inspired by Bill's quiet passion and rigorous professionalism. He complained about "not much liking this teaching business," referred to writing exams as "a dreadful task," and spoke of having "survived" the chaos of course registration, but he also mesmerized undergraduates in the natural history class by imitating a heron, his crooked right arm jabbing at imaginary prey. One day he passed around a live glass frog, so named because its green bones and beating heart are visible through the translucent belly skin, and deadpanned, "This one doesn't have many secrets." On a field trip he brushed aside concerns about a threatening storm with "We'll get these kids wet and exhausted hunting salamanders. They'll love it!" He stretched us grad students, too, into more critical ways of thinking. We read Darwin's *On the Origin of Species* for a seminar course,[18] and years later I realized Bill's reticence encouraged shier members of our group to speak out, a lesson that would restrain my own commentaries as a professor at Berkeley.

Bill's colleague Tom Kennerly, who'd studied gophers with Frank Blair at Austin, also excelled at field teaching. His attention to trapping and preparing mammal specimens would have made M.V.Z.'s Grinnell proud, and on class trips he gestured at roadside habitats from a weaving van as if they were all that mattered. To better appreciate "pioneer organisms," at a local cemetery we inspected twenty-five tombstones in each of four age categories. Polished new ones were barren, but on those only a few years old we encountered lichens, the symbiosis of rock- and bark-dissolving fungi with photosynthetic algae or bacteria. Still older markers had lichen growth as well as moss on their most favorable corners, and nineteenth-century gravestones, their tops and sides converted to soil, hosted flowering plants within lush coverings of more primitive organisms. Tom had cleverly shown us a century of ecological succession in an afternoon.

For a thesis "defense," the student presents research and answers questions; then the professors, if satisfied, formally approve the document. Mine commenced with a routine summary of results and implications, beginning with an ethogram of twenty-eight coralsnake behaviors and the experiments on cues controlling headfirst ingestion. Museum diet samples showed that the feeding biology of one species was fairly constant from Florida to Texas, despite size-related, seasonal, and geographic variation. Finally, I'd studied nearly half of the other species as well as their close Asian relatives, and with few exceptions—one Amazonian species ate fish, another velvet worms—a diet of snakes and other elongate vertebrates typified the group.

Things went well until a new assistant professor showed up and demanded justification for my thesis topic. "Because coralsnakes are beautiful and fascinating," I replied, caught off guard, and Robert McMahon shot back, "You mean this is just a *game* for you?" I sat speechless as Bill jumped in with "Oh come on, Bob, during physiological experiments on snails you're thinking, 'I'll study mollusks and save *humanity?*'" Many years later I would be embroiled in a parallel debate about the importance of natural history and see wisdom in Bob's question, but that afternoon I was deeply grateful for my advisor's protective response. The few minutes I waited for the committee's favorable decision seemed interminable. The sixty-six-page thesis was the longest document I'd ever typed.

McMahon's challenge notwithstanding, two books I read at U.T.A. underscored the interplay of natural history and scientific theory, and a third solidified my career path. George Schaller's *The Serengeti Lion* summarized three thousand hours of fieldwork in service of questions like, how do carnivores affect prey populations and why does that species live in groups, when most cats are solitary hunters?[19] Peter Klopfer's *Behavioral Aspects of Ecology* linked what individual animals do with antipredator adaptations, causes of high tropical diversity, and other classic puzzles, while R. F. Ewer's *Ethology of Mammals* showed how studying diverse behaviors across a group of animals might link my interests in other aspects of biology.[20] Accordingly, and having noticed that academic positions were usually advertised for evolution and other conceptual disciplines rather than taxonomic specialties like herpetology, I sought a Ph.D. program in which to study reptiles from the standpoint of issues framed by these authors.

Gordon Burghardt's publications on reptile behavior had caught my attention while Donna and I lived in Germany, and we'd taken his "readers' advice," in a European travel guide, about renting motor bikes in southern France. So after some correspondence about mutual interests, I gave a talk about coralsnake feeding to his research group at the University of Tennessee. The zoology department's admissions committee discounted my earlier bad grades in the face of respectable Graduate Record Exam scores and publications, and thus late that summer, newly divorced, I was off to the Volunteer State.

Soon enough I'd endure a doctoral exam that made Bob McMahon's questions seem cheerfully supportive.

In 1973 the Nobel Prize in Physiology and Medicine, an honor previously unimaginable for organismal biologists, went to Niko Tinbergen, Konrad Lorenz, and Karl von Frisch for founding ethology. Burghardt was beaming about the laureates when I arrived in his Knoxville lab that fall, and I couldn't have been prouder of my newly adopted discipline or more awed by my advisor.[21] Only four years older than me, he was a widely read, creative thinker and committed Darwinian. Right away I noticed a quote on his desk from Norman Maclean, a University of Chicago English professor who years later would shape my own literary ambitions: "A good teacher is a tough guy who cares deeply about something that is hard to understand."

Reptiles had fascinated Gordon since his Wisconsin childhood, when he'd bought green anoles from a circus, kept red-eared sliders as pets, and caught gartersnakes in vacant lots. Initially a chemistry major at Chicago, he switched to biopsychology and fell under the spell of an austere but inspiring professor. Eckhard Hess pioneered studies of imprinting—by which young animals form long-term attractions to specific stimuli, often associated with food or other animals—and unlike most psychologists at the time he espoused ethological perspectives. Hess gave Burghardt lab space for undergraduate research on turtles and lizards but couldn't allow snakes, because his wife hated them. Gordon crammed cages of newborn gartersnakes into his tiny student apartment, and only after Hess encouraged him to stay on for a doctorate did the two of them overturn the "no snakes" rule.

Gordon's dissertation combined his knack for chemistry with an interest in young animals, addressing Tinbergen's four questions in terms of how naive young of ecologically diverse serpents respond to prey cues. Plains gartersnakes, for example, attacked cotton swabs laced with the scent of leeches, worms, fish, or frogs, but they only tongue-flicked cricket and mouse odors. Other species also preferred chemical signatures of their natural diets, and baby queen snakes reacted only to *freshly molted* crayfish, consistent with specialization on soft-shelled crustaceans. Exceptions were instructive too: Butler's gartersnakes eat leeches and worms, but neonates also responded to fish and frog odors, so their preferences evidently hadn't yet diverged from those of close relatives; corn snakes and cottonmouths, which have broad diets, scarcely discriminated among extracts, suggesting that for them preprogramming might be detrimental. Overall, snake behavior, evoked by specific stimuli and without prior experience, nicely matched ethological notions of instinct.

When we first met, Gordon had thinning brown hair and a moustache that later morphed into a beard. My preconceptions about lab-bound psychologists, preoccupied with caged pigeons and white rats, were soon squashed by his enthusiasm for a battered old Jeep and snake hunting in the nearby Cumberland Mountains. He was proud of a summer spent at Lorenz's research institute in Bavaria, and his contagious enthusiasm for all things ethological fostered a challenging intellectual atmosphere. Gordon gave talks at meetings and openly grappled with theoretical issues, for example, and he eagerly tried novel approaches, from tropical fieldwork to high-tech brain imaging. Besides snakes, he was studying black bears in the Smoky Mountains and green iguanas in Panama. And down the road, as we will ponder later, he would broaden Tinbergen's manifesto to include a bold fifth question about the inner worlds of animals.

Gordon dealt praise sparingly but was otherwise supportive, even cosigning a bank loan so I could buy a car, and we graduate students admired our professor all the more for a certain quirkiness. Among his eccentricities was a fondness for Honduran cigars, smelly brown things that everyone dreaded him lighting up. At one social gathering, shortly after lecturing to our advanced ethology class about sign stimuli and their corresponding internal releasing mechanisms—witness the attractive effects of a male stickleback fish's red belly on females of the same species—he fired up a huge stogie and placed it on an ashtray under his chair. Within minutes a Siamese cat sauntered up and made stereotypical fecal burying movements, at which point Gordon sheepishly laughed with the rest of us about its ethological commentary on those cigars!

For a course project that first semester I videotaped captive boas and pythons, then confirmed with slow-motion film analysis that they applied killing coils in identical fashion, yet differently from more recently evolved ratsnakes and kingsnakes. My term paper explained that while homology, a core concept in comparative anatomy, refers to ancestral resemblance—the homologously enlarged incisors with which rats and squirrels gnaw food, for example, date back to the ancestor of all living and extinct rodents—convergence, in contrast, is independently evolved similarity, often associated with a specific function and presumably favored by natural selection. Wings of birds and bats are thus homologous *as forelimbs* and convergent *as wings,* given that their respective closest relatives are reptiles and mammals with front legs that have not been modified into flight structures. The common ancestor of birds, bats, and insects was a far more ancient, simpler organism that lacked appendages of any sort; hence, neither the wings nor limbs of insects are homologous with those of vertebrates.

Homology thus reflects biological *heritage,* manifested as inherited similarities among organisms, and if the patterns I'd discovered held up to more extensive taxonomic sampling, identical prey-killing modes of boas and pythons implied descent from a common ancestor. Moreover, since fossils showed that that ancestor lived at least ninety million years ago, the homologous behavior of extant constrictors had to be that old too. I was feeling enthusiastic about these preliminary results until someone in class asked whether anyone else would care—or in more practical terms, why should a granting agency fund this research? My rejoinder was that textbooks bemoaned the lack of behavioral fossils, other than occasional dinosaur trackways and mosasaur bite marks on ammonites, so comparisons among surviving species might provide new data on such things. Here was a way, I hoped, of measuring the geologic age of behavior.[22]

Gordon scribbled suggestions on the term paper and pointed out a recently published critique that raised the intellectual stakes for my project. I had just assumed one could study what animals do in terms of evolutionary history, as comparative anatomists did with bones and Lorenz had done with the social displays of ducks and geese; I didn't know that ichthyologist James Atz had asserted that homology was irrelevant because behavior, unlike anatomy, he claimed, is highly variable and easily modified by experience.[23] Moreover, Atz asserted, behavior is strongly influenced by selection, so convergence likely explains similarities among all but the most closely related species. I'd already seen, though, that tail displays and a diet of elongate vertebrates characterized dozens of Old and New World coralsnakes, and now I found the same constricting behavior in several major snake lineages. Since common ancestry was a straightforward explanation for those patterns, Atz's sweeping criticisms seemed wrong-headed, and I began developing a dissertation proposal along those lines even as I set about satisfying other Ph.D. requirements.

The two-part qualifying exam was a disaster, thanks to my increasingly overt self-confidence and the peculiar culture of graduate education. On five successive mornings I typed out answers to my professors' written queries, which I found challenging but well within my grasp. A week later, however, the oral component began with a physiologist who grumbled, "This isn't fair because you dropped my course, but what are Herbst corpuscles?" After it was clear I was ignorant of far more than sense organs in bird knees, Susan Riechert, my zoology department coadvisor whose in-class drawing of a sharklike flatworm mouth I'd corrected during our first semester, barraged me with hard-hitting ecology questions. When she finished, my head was pounding, my stomach in knots. I couldn't recall simple details of biochemistry, and over the course of three hours things declined to a humiliating finale. For the "writtens," Gordon had asked me to discuss six of twelve pioneer ethologists; I hadn't subsequently looked up the others, so the ordeal ended as I slunk out the door with him muttering, "Well, you don't know much about that either, do you?" I barely passed and to this day always remember that bewildering experience when serving on doctoral exams myself.

The dissertation developed into an appraisal of behavioral homology, backed by a survey of constriction and defense in snakes. I countered Atz's skepticism by observing that antlers and some other structures are also variable, yet conversely many action patterns are as stereotyped as bone shapes; moreover, newborn animals often exhibit identical behaviors across taxonomic groups. Clearly anatomy isn't fundamentally unique in those respects, so I would address serpentine feeding and antipredator tactics with a question that tantalized Darwin, Wallace, and Fitch: Why are there similarities and differences among species? Since extensive taxonomic sampling in the field would be impossible given the time constraints of a Ph.D. program, I videotaped constricting behavior by snakes in the Atlanta, Dallas, Fort Worth, Houston, and Knoxville zoos. I also observed captives in Gordon's research lab, including an exceedingly rare Oaxacan dwarf boa sent from Mexico by my friend Jon Campbell, by then a graduate student with Bill Pyburn.

Almost six hundred observations of seventy-five species showed that constrictors applied loops around prey in nineteen ways, but forty-eight relatively primitive species always used only one of them. Baby pipesnakes, boas, and pythons employed the exact same movements too—no experience required; furthermore, their ancient style persisted in the face of ecological diversification. Green tree pythons, giant aquatic green anacondas, and the secretive Oaxacan dwarf boa, despite their diverse habitat preferences, all subdued prey with identical coils, which I proposed had facilitated the origin of a large gape in some snakes (I'll cover this topic in more detail in chapter 9). Conversely, antipredator mechanisms, descriptions of which I gleaned from the natural history literature and my own field studies, have generally evolved more recently and are fine-tuned to local ecological conditions. One hundred and twenty-four species for which I had observations defended themselves with tactics that are more reliably predicted by lifestyle than legacy; arboreal species the world over have independently developed open-mouth threats, for example, and unrelated burrowers use tail displays.[24]

More generally, I'd confirmed the early ethologists' claim that comparative studies of behavior could be fruitful, and the final defense went off without a hitch. It would be more than a decade, however, before I began to reflect on how behavioral homology might inspire nature appreciation.

Those four years at Tennessee contrasted sharply with the orderliness and normality that characterized my youth. I shared a four-room farmhouse twenty miles north of Knoxville with my second wife, Dona (we'd married soon after I moved to Knoxville), and two venomous beaded lizards, two tame opossums, nine cats, a dog, and an African pygmy goat. Sluggo had vacant golden eyes, spiraled horns, a typical goat penchant for licking his lavender erection whenever visitors approached our porch, and a resolve to kill Layla, the Malamute, by slamming her against the house. The larger beaded lizard was an escape artist; on one nocturnal foray he crawled into a hole in the bathroom wall, and I had to tear out a piece of plywood behind the sink to retrieve him. One early morning I awoke with sudden fears of a heart attack, only to find myself nose to nose with fourteen-pound Posey the possum, asleep on my chest. Throughout that craziness, biology was more and more my life, though I was still figuring out just what it all might mean.

Scientifically, my chief role models were Fitch as natural historian, Pyburn, Burghardt, and Riechert as intellectuals. Bold theoretical forays in Evelyn Hutchinson's *The Ecological Theater and the Evolutionary Play* and Robert MacArthur's *Geographical Ecology* encouraged me to seek broader implications of snake biology, as did Dan Janzen's "Why Fruits Rot, Seeds Mold, and Meat Spoils," as well as his other clever papers in the *American Naturalist.*[25] Thomas Kuhn's *Structure of Scientific Revolutions* and Karl Popper's *Conjectures and Refutations* emphasized the importance of skepticism and evidence, and I began developing a feel for scholarship by, for example, comparing Emmett Dunn's thoughtful analyses of snake evolution with bitter rival Ed Taylor's careless description of male and female frogs from the same population as separate new species.[26] Along the way Mary Ann Handel, my favorite "cell smasher" at U.T., and Sandy Echternacht, our new herpetologist, exemplified goal-setting, careful preparation, and concern for students as key attributes of good teaching.

Students are often sobered to find academia shaped by personal frailties and politics as well as by its much-vaunted search for truth. We slogged through a rigorous biometry course, but when I challenged a visiting molecular biologist's lack of statistics one of my professors admonished me, "This guy's so good he doesn't need them." We struggled with public speaking in the face of withering criticism from our teachers and each other, then were astonished by the sloppiness of another renowned guest, a disheveled ecosystem theorist who scarcely paused as his overhead transparencies slid to the floor. As for the intense self-focus and ego-jostling that comes with grad studies, antidotes came from the likes of Leonard Radinsky, a lanky, shaggy-haired New Yorker who studied tapir evolution for his Yale Ph.D., then pioneered fossil brain research at the University of Chicago. I'd sought his advice on the anatomical basis for behavior and asked, with obvious envy, "What's it like to be famous?" Len countered, "How many people do you think have heard of me?" After a moment I guessed several hundred, whereupon he said with a grin, "Uh huh, so what's the big deal?"

Midway through my Ph.D. a blockbuster book by Harvard's Edward Wilson drew me into controversy. *Sociobiology* provoked a debate on genes and human destiny, exemplified by Peter Klopfer asserting that behavior, unlike anatomy, doesn't have a genetic basis and cannot evolve.[27] Flaws in the Duke professor's arguments paralleled those in James Atz's homology critique, serving as a foil for my dissertation as well as the subject for my oral presentation at an Animal Behavior Society meeting held at the American Museum of Natural History. When several dozen snake species, including their naive young, kill prey exactly the same way, I asked in my talk, why not turn to evolutionary heritage as a parsimonious explanation? I was blindsided, though, when Ethel Tobach, the program host, accused me of falsely maligning Klopfer; fortunately, Burghardt caught my eye from the sidelines with a nod I took to mean, "Be gracious." Tobach was Atz's colleague at the A.M.N.H. and evidently resented ethological perspectives because of Konrad Lorenz's wartime association with Nazi ideology,[28] but during a coffee break she recanted her criticism of me when confronted with Klopfer's published essays.

Wilson's *Sociobiology* also fueled my passion for conservation, by way of his unabashed celebration of biodiversity. I was understandably focused on the special plight of snakes, and one day an office mate mentioned that some students investigating black bears in the Smoky Mountains had made hatbands out of the skin of a timber rattler, supposedly slaughtered for the sake of their own safety. I fired off an angry letter to the local newspaper, outraged that fellow scientists would kill any animal in the national park and rhetorically threatening to shoot bears there in preemptive self-defense. Gordon's phone was soon ringing, and I was grateful for his unwavering support in the face of heavy pressure from the snake killers' professor to censure me.

Although Tennessee was a good fit for grad school, reptiles were scarce, and I craved tropical fieldwork. My first semester, after Gordon talked glowingly of Panama, I offered to borrow plane fare if he'd take me with him the next time, which he did, using research funds. We would spend three field seasons on Barro Colorado Island, which was owned by the Smithsonian Tropical Research Institute and closer to idyllic than anything I'd ever imagined. Amenities included rustic lodging and family-style cuisine, but no telephones or televisions. The forest harbored abundant wildlife, including howler and spider monkeys, my first wild primates. Large black and tan weasels called tayras bounded through the lab clearing, and American crocodiles prowled the surrounding inlets. Harmless seven-foot-long bird snakes inflated their necks and gaped when I grabbed them, whereas the smaller but deadly Central American coralsnakes thrashed about with tails curled up and elevated, ready to bite anyone attacking their fake "heads."

With Gordon and another student, Beverly Dugan, I spent hundreds of steamy hours at a clearing on Slothia, a tiny islet off Barro Colorado Island in Gatun Lake, on which dozens of green iguanas, several red-eared slider turtles, and an American crocodile laid eggs. S.T.R.I. scientist Stanley Rand had already filmed this reptile rookery, including iguanas fighting for nest burrows and a huge female croc appearing suddenly in front of his flimsy observation blind.[29] Stan's talk at a herp meeting resonated with my advisor's long-standing dream of studying newly hatched reptiles in nature, and on his pilot visit Gordon was encouraged by the hundreds of baby turtles and lizards emerging from a few square yards of exposed soil.

Our blind was a tiny wooden shed, open from behind, in which we sat on stools and poked telephoto lenses through slits in a burlap front wall. We arrived in early May, just before the onset of the rainy season, when eggs hatched. That first year we discovered sociality previously unknown in reptiles other than dinosaurs: Groups of ten-inch green lizards emerged from nest holes, swam across open water, and dodged attacking birds. They munched flowers within inches of each other without aggression. One head-bobbed at siblings, ran several yards, looked back, returned to the stragglers, head-bobbed again, and seemingly led them to nearby vegetation. We found iguanas sleeping together in bushes at night and marked them with paint spots, then by observations confirmed that they spent 98 percent of daytime hours immobile. Nonetheless, the babies were surprisingly cognizant, cocking their heads as bananaquits and other small birds flew by, and feeding when wind rustling the vegetation concealed their movements.[30]

Caught up in natural history, without a clue to the hows and whys of it all, I inevitably had moments of introspection and inspiration. Once as we assembled for lunch Ron Carroll, an ant ecologist, reported a brown vine-snake eating a baby iguana nearby. "They specialize on anoles," I told him with overconfident authority, "but I'll have a look." "Well it *is* eating an *iguana*," Ron replied with a smile, and I managed to laugh at myself while watching our paint-marked lizard head down the serpent's gullet. Another day as I paddled a canoe with Cathy Toft, a Princeton grad student studying frogs, we photographed a Baird's tapir in a shoreline "Eo-scene" that later inspired me to think of living creatures as icons for their extinct relatives. That tapir, nicknamed Alice, often entered the lab clearing for handouts, so one night, suddenly curious about what she felt like, I snuck up from behind. Alice's ample derrière brought to mind a warm, stubble-covered watermelon, and later I fell asleep marveling at her ancient alien strangeness.

Burghardt's other Ph.D. students, an eclectic bunch whose intellectual breadth and future paths were as diverse as their research, also taught me a lot. Bev Dugan used hillside vantage points to untangle social conventions among iguanas; she would later combine experimental perspectives with exceptional people skills in a successful management career. Hugh Drummond, among the first to study snake foraging in the field, subsequently turned his attention to siblicide in sea birds when he returned to Mexico. Doris Gove stayed on in Knoxville, where she parlayed research on northern watersnakes into children's books on local natural history; Elizabeth Shull applied the ethological principles she'd learned studying mockingbirds to veterinary medicine. Paul Weldon has most closely emulated Gordon with a career in chemical ecology, while Hal Herzog, who initially investigated maternal behavior in alligators, and then cockfighting as a cultural phenomenon, has recently published *Some We Love, Some We Hate, Some We Eat*,[31] a brilliantly insightful book on human-animal relationships.

Our friendships unfolded in many a beer-soaked, horizon-expanding conversation. Bev insisted I read David Halberstam's *The Best and the Brightest*,[32] which laid bare the Vietnam conflict as political folly and shook me badly.[33] Then, provoked by a quote that ended Wilson's *Sociobiology*, I plowed through the writings of Albert Camus. For years I'd been vaguely aware of an emotional load—Bob Dylan's *Blood on the Tracks* was inexplicably disturbing, and fumbling to express my resentment of the war to another student, I surprised us both by choking up—but the French writer's insights on alienation cracked open the gloomier sheds of my psyche and struck a match. Mortality, I began to see, was still cloistered for most people I knew, death acknowledged in somber rituals that carried little of the sights or sounds of tragedy; yet as a youth I'd touched a woman's most intimate places, barehanded in those days before AIDS, to deliver her baby. I'd seen splattered skulls and shrieking survivors, watched a grief-stricken mother shaking so hard I feared she'd eviscerate, and learned firsthand how twisted minds could turn lust and love into murderous rage.

By 1977 I'd packed a ton of discovery into two graduate degrees and been in Knoxville longer than anywhere else in my life. Snakes were my organisms of choice, evolution and ecology my conceptual foci. Education and conservation beckoned as well, and my emotional lid, long screwed down, had ever so slightly loosened. In the spring I had a disastrous interview at Cornell. That fall I took a position teaching animal behavior at the University of Pennsylvania's College of Veterinary Medicine, but its emphasis on domestic species clashed with my goals, so the following summer I moved on to Berkeley. For the next couple of decades I would roam hot dry places and explore hot wet ones, preoccupied with studying nature, yet learning more about myself in the process.

SEVEN

Hot Dry Places

THE MOHAVE ENCOMPASSES TWENTY-FIVE THOUSAND square miles of southern California and adjacent states, threatening to bake anyone who dares enter. By June it's an oven and in August, even after dark, a merciless furnace. Scramble up a boulder on a breezy April morning, though—say, on a flank of the Granite Mountains—and you'll bask in sumptuous austerity. Snow-capped San Jacinto pierces azure sky a hundred miles to the southwest, ten thousand feet above the Los Angeles Basin's smoggy border. Joshua trees dominate nearby rocky slopes. Purple flowers carpet the flats. A friend's boots crunch with metronomic cadence from an arroyo below, and the buzzing of a pesky fly feels like starlight, crisp but vanishingly small. Everything is weathered. The air *tastes* clean. As afternoon shadows deepen, distant ranges turn flat lavender gray. Creatures that perish here dry quickly, minus the tariffs of scavengers, and drift off on the wind.

Like other deserts—Great Basin, Sonoran, and Chihuahuan in North America; Sahara, Taklimakan, and so forth elsewhere—the Mohave is defined by climate. From overhead in a jet it looks like elephant skin, etched with dry washes that merge into empty riverbeds, their fractal tendrils dotted with shrubs and occasional trees. On the ground, out walking, questions come to mind for which answers involve water, wind, and sun: Why are these leaves leathery? Why are so few mammals out during the day? Nowhere else is our molten yellow ball so omnipotent, or moisture so precious. And surely no other night skies so captivate us. Deserts *feel* cosmic, so it's not surprising that religions spring up in them, that they inspired Rumi's mystical poetry and Camus's spare philosophical prose. Arid lands were Henry Fitch's favorite habitat too, because, as he told me with characteristic economy, "They're open and full of interesting plants and animals."

For now, a giant Mohave succulent and two small creatures with almost identical names can underscore why herpetologists are drawn to deserts. Joshua trees are actually yuccas and, like agaves, palms, and other monocots, don't make wood or bark. They often occur in clusters of fibrous trunks, upwards of fifteen feet tall and variously bent, as if a troupe of gangly dancers with Afro hairdos froze in mid-routine; their tough, daggerlike leaves blast out in eighteen-inch rosettes, collectively resembling the crowns of real trees and adorned by creamy-white, moth-pollinated flowers that smell like gorgonzola cheese. Fallen Joshua trees foster a moist duff favored by our quarry, though searching requires caution—their trunks and branches are covered with sharp leaf scars, their underlying microhabitat easily disrupted, and within them big scary arachnids occupy the same nooks and crannies preferred by reptiles.

Desert night lizards, often common under yucca logs, are no larger than a stout wooden match, with tiny legs, minuscule gray-green scales, vertical pupils, and gogglelike spectacles instead of eyelids. I've also caught them beneath trash in barren scrub, with no Joshua trees in sight, while one was sighted among creosote bush roots; so they might be ubiquitous across the Mohave, numbering in the gazillions. As do many other lizards, these gecko look-alikes eat arthropods, but unlike others here, they live in families and give birth to young rather than lay eggs.[1] Desert nightsnakes, found less often, are the size of a big pencil, with flattened heads, dark-spotted bodies, and pearly bellies; rear-fanged yet easygoing when handled, they subdue lizard prey with a venomous bite. They are related to frog-eating tropical cat-eyed snakes, and likewise crawl about at night, hiding under surface objects by day.

Given the zeal with which captive desert nightsnakes consume desert night lizards, herpetologists long assumed a deadly intimacy between the two reptiles. My students and I were thus surprised to discover mostly diurnal species in the stomachs of these little serpents: of ninety-two food items in museum specimens, only four were nocturnal lizards. This paradox, that "night" snakes mostly eat "day" prey, was resolved during a conference in Baja California when I skipped lunch to scout nearby arroyos. While strolling about I lifted a board; wary of scorpions, I discovered instead a nightsnake swallowing an orange-throated whiptail half its size. Because there were no burrows under the board from which the predator could have stalked a sheltered lizard, it must have arrived earlier to ambush a wide-foraging prey. Confirmation of this scenario came on another afternoon that same spring. An M.V.Z. colleague saw a side-blotched lizard enter a hole and emerge

Desert nightsnake, Pajarito Mountains, Santa Cruz County, Arizona. Despite its name, this little serpent feeds mainly on diurnal lizards. (Photo: H. W. Greene)

struggling, whereupon he pulled it free with a nightsnake attached to one leg! Together our observations trumped conventional wisdom, by showing that these temperate-zone "cat-eyes" are diurnal stationary hunters.[2]

So Fitch was right: arid landscapes *are* fascinating, and maybe I'd imprinted on them as a child, like the orphan goslings that followed ethologist Konrad Lorenz's every move. I attended first grade in Tucson, where even today Sonoran Desert Gila monsters tumble into backyard pools and coyotes take the occasional house cat. Kangaroo rats kicking sand at a sidewinder rattlesnake in Disney's film *The Living Desert* also captivated my youthful psyche. And like Henry, as an army medic stationed in El Paso I coped with wartime uncertainties by poking around Chihuahuan Desert bajadas. Then in 1971, freshly discharged from the military, I returned with fellow grad student Jerry Glidewell. We sought Texas alligator lizards in Laguna Meadow and marveled over a butterscotch-colored Trans-Pecos ratsnake on the road near Terlingua. We braked hard for a mountain lion in Panther Pass—"It's him, it's him!" Jerry exclaimed—and identified chunks of glossy snake in a road-killed badger's stomach on the plains near Marfa.

The place was smacking me with priceless moments and I wanted more. Back then Arlington was a suburb of Fort Worth, nicknamed "where the West begins," and the poor sister of glitzy Dallas, thirty miles to the east.

What is now one sprawling metroplex straddles a midcontinental ecotone stretching from Manitoba to Mexico, along which forest remnants mix with what once was prairie all the way to the western cordilleras. The following spring Bill Pyburn's herpetology class visited the Eagle Mountains, an outlier of the Sierra Madre Oriental one hundred miles southeast of El Paso. On the ten-hour drive we'd first passed from lush Trinity River bottoms and oak-hickory savannas into more open country, from trees and moisture surplus to grass and moisture deficit. Next our van traversed a monotony of ranchlands speckled with oil derricks and the occasional refinery, then finally crossed the Pecos River into Chihuahuan Desert proper. I was daydreaming about reptiles when Pyburn squinted off to the left, noted the Davis Mountains' dark jagged profile, and allowed we'd make camp in time for sunset over the Eagles.

In a photo from that trip I'm smiling from under a straight-brimmed miner's hat and Pancho Villa–style mustache. Crimson stains dominate one pant leg because that first night in the Eagles, distracted by an enormous silver moon, I'd collided with a lechuguilla agave and punctured my shin. For breakfast we washed down Wanda Pyburn's *huevos rancheros* with cowboy coffee, then wandered about as worldly cares faded with the morning shadows. Birds sang from bushes, lizards patrolled cobble piles, and most other creatures were avoiding the sun. Six-inch centipedes coiled with their eggs beneath rocky slabs, as if guarding precious jewels. Black-necked gartersnakes hunkered under boards beside a drying pond, implying that the frogs they eat were thereabouts too. Glidewell came in for lunch with a venomous Chihuahuan lyresnake he'd extracted from a crevice and broke out grinning when Bill admitted never having seen that secretive species. Around a camp-fire, after dark, we debated careers in science and cursed the wildlife biologist we'd met driving in who bragged of killing rattlers.

West Texas got me thinking about how individual lives relate to species' distributions, a problem that has long fascinated ecologists. Those forays, however, were followed by a hiatus in which I earned two advanced degrees, landed an academic job, and not once set foot in any desert. My master's thesis and Ph.D. dissertation were lab based; the only fieldwork I did was when studying Mexican snakes and Panamanian iguanas. Then in 1977, fresh out of graduate school, I declined a Smithsonian postdoctoral fellowship to teach at Penn's vet school, only to discover I was ignorant of horse innards, student interest in ethology was limited to unruly pets, and Philadelphia felt too *eastern*. The next spring I interviewed for retiring M.V.Z. curator Robert

Stebbins's position and in August, fulfilling a childhood dream of becoming a herpetologist, moved to Berkeley.[3]

My desert yearnings were all the more powerful for having been confined to buildings and forests for five years. Struggling up Appalachian gorges I'd found salamanders galore but reptiles few and far between; blue sky appeared through foliage openings, as if life were confined to deep concavities, but I craved space so far in all directions that Earth's convexity was palpable. I just couldn't forget scraggly horizons, hinting of new discoveries yet proclaiming, "Don't get cocky. Even the rocks aren't permanent." I knew that on broiling summer afternoons most of the Eagle Mountains' forty-five species of reptiles would be hidden, as if ordered inside until visitors left. Deserts, it seemed, offered a glimpse of life's secrets for those who looked hard enough, and luckily, I'd soon be within an easy day's drive of one.

Serving on a job search committee during grad school paved my way to California—with a detour in the wrong direction. Tennessee had advertised for a herpetological ecologist and got some hundred applicants, a few of them blatantly inappropriate—"I study ticks," wrote one, "but like snakes"—and others without noteworthy accomplishments. Of about twenty taken seriously, five interviewed, and Sandy Echternacht was hired in time for me to TA his herpetology course. During the process I noticed how candidates' talents shaped perceptions of departmental needs—"Hey, this one could teach biogeography!"—and concluded that my own success would depend on applying to lots of places and consistently being among the top contenders. Above all, I realized that the most promising applicants stood out for getting grants and publishing, not for where they'd studied. That spring I tried for a dozen positions, had a couple of interviews, and landed at Penn.

Months later I wasn't optimistic about escaping Philadelphia, because competition for the Berkeley position was tough and others had stronger museum credentials. I'd collected lots of specimens but didn't mainly describe variation and discover new species, so I figured the interview would be merely a chance to meet people and see cool herps. As it happened, after the usual formalities, I was sent on a class field trip, during which the TA was to observe my teaching skills and report back to mammalogist James Patton, chair of the search. Briones Regional Park teemed with animals known to me only from books and Henry Fitch's early papers, and I couldn't resist extolling a southern alligator lizard's prehensile tail and soft-skinned side pleats.

"Watch him puff up," I gushed. "Why else might something in bony scales expand? 'Food'? 'Eggs'? Good answers!" I also persuaded a student to sniff the red, coiled tail of the ring-necked snake she'd found to better appreciate its defenses. All of which must have impressed my hosts, and within months I moved to the West Coast. Among the perks of my new job was founding director Joseph Grinnell's antique rolltop desk.

Over the next two decades I often figuratively pinched myself in disbelief. For starters, the University of California's oldest campus is heartbreakingly beautiful, sprawled through oak woodland and chaparral in Strawberry Canyon and looking out over San Francisco Bay. Mountain lions and golden eagles still patrolled its fringes, as they had in Henry's grad school days, thanks to the extensive system of East Bay Regional Parks. Botanical Garden signage warned of rattlesnakes, and one of my first official tasks, after being summoned by campus police, was capturing one on the rock wall of a child-care center by the football stadium. Shortly thereafter, I witnessed a poignant nod to history in the canyon's redwood grove: as family and friends memorialized Carl Koford, pioneer condor biologist and among Grinnell's last Ph.D. students, a circling red-tailed hawk graced us with its descending scream, as if nature's bugler were playing "Taps."

Although Berkeley proved as magical as Henry remembered, the intellectual scene was as daunting as it was exhilarating. The search committee had perceived me as most enthusiastic among the interviewees about teaching natural history, and mine was the only application that emphasized a commitment to conservation, strengths of Bob Stebbins's that they believed worthy of continued emphasis. Fair enough, I thought, but the zoology department, wherein M.V.Z. curators held professorial appointments and were judged for promotion, brimmed with giants. During faculty meetings I'd likely as not sit between endocrinologist Howard Bern, an award-winning teacher and member of the National Academy of Sciences, and embryologist Richard Eakin, featured in *Life* magazine for lecturing in costume as Darwin, Mendel, and other famous figures.[4] On the junior end of the ranks, ecologist Wayne Sousa arrived the year before me, so promising that he was hired fresh out of grad school, and Mimi Koehl, already a rising star in biomechanics, came a year later.

Four professors became dear friends and helped balance the demands of my new job. Marvalee Wake studies caecilians, once-obscure amphibians that thanks to her now show up in the most prestigious scientific journals. She also advocates quality teaching as well as broader service, and was a superbly fair, effective departmental chair. David Wake, whose creatures of choice are salamanders, owns the most profound feel for the development, internal workings, and evolution of organisms I've ever encountered. As M.V.Z. director he was also my boss, and we had common goals as well as amiable disagreements that still challenge me. Mammalogy curator Jim Patton's passion is rodent evolution, and he always led our team-taught natural history class up the steepest hills, followed by panting students and staff. Over the years Jim looked more and more like Grinnell, whereas I resembled a younger Patton—and thus Grinnell—a few years behind them both in hair loss and beard style. Rob Colwell, who studied nectar-eating mites that travel among flowers in hummingbirds' nostrils, nurtured my community ecology interests and was always ready with wise advice when I most needed it. Each in their own ways, Marvalee, David, Jim, and Rob were my most important professional role models.

I launched my biennial spring herpetology course in that heady atmosphere, closing in on a teenage dream and confident of succeeding at least as an instructor. We'd use conventional lecture, lab, and field formats, teaching facts and theories as well as immersing students in discovery. I jazzed up lectures with topics not in our text or that cried out for research, striving to portray a dynamic science. Marvalee volunteered her stunning slides of caecilian eyes, hidden under skull bones, and of the glands with which those burrowing creatures feed developing young on fatty "uterine milk." David taught me to liken the motion of salamanders shooting down insects to that of squeezing watermelon seeds between fingertips—their muscles project the slippery Y-shaped throat skeleton, its forward-facing base armed with a sticky tongue pad for prey capture. I also sought aids from colleagues elsewhere, like raptor biologist Stan Temple's photos of Round Island skinks cracking bird eggs by rolling them off rocks, then eating the contents.

My TAs, themselves accomplished herpetologists, familiarized undergrads with diversity at local to global levels. Students watched captive rattlesnakes to safely differentiate species using color patterns; they hunkered over microscopes to distinguish frogs from all over the world on the basis of pectoral girdles, tadpole mouthparts, and other anatomical peculiarities. Twice during the semester we administered "practical exams"—preserved red-bellied newts, for example, were accompanied by questions regarding identification, sex, distribution, and predator avoidance. Together lectures, readings, labs, and tests provided a scaffolding of facts and concepts that would make more sense when the class went looking for salamanders in creeks and lizards in sand dunes.

Our field trips took advantage of California's extremes: cool and moist in the north, hot and dry in the south, with emphasis on amphibians in the former and reptiles in the latter. Pretrip handouts laid out routes and schedules, emphasized safety, and listed things to bring: shorts and pants, long-sleeved shirt, hat, bandana for dust storms, jacket, sunscreen and sunglasses, boots, water bottles, binoculars, camera, notebook, sleeping bag, eating utensils, and so forth. M.V.Z. supplied stoves, pots, pans, and ice coolers, and everyone chipped in for groceries. Students were responsible for cooking and cleanup, with instructions that I expected plenty of tasty food and good coffee! There were humorous interludes, of course, like when we were calculating how much toilet paper to buy and one guy bragged about making do with two squares per "event"—whereupon the gal next to him muttered, "You are *not* cooking!"

The course's shakedown cruise, to Mendocino County, commenced on a Friday in March, with Saturday's sunrise revealing frost on our tents. Otters prowled the Eel River by camp, and nearby rocks harbored western fence lizards and a western skink or two, although once we uncovered only a suspiciously fat, brown-and-yellow-banded California kingsnake. Ranging over grassy, oak-dotted hillsides, we palpated dusky-footed woodrats out of western rattlesnakes and counted dozens of black salamanders under bark and rocks. Wading chilly streams, dwarfed by massive old-growth conifers, my students inverted their binoculars to magnify the gills of torrent salamander larvae and a tailed frog's mating appendage. Flipping planks in a meadow yielded readily identifiable northern and southern alligator lizards as well as three confusingly similar species of gartersnakes. And with rough-skinned newts in hand we inspected the male's slimy skin, tail fin, and swollen feet, temporary transformations for aquatic mating in an otherwise terrestrial amphibian.

Meals really hit the spot after hiking such breathtaking locales, and later I'd drift off to sleep as students joked and sang around a bonfire. Driving home Sunday we'd picnic beside a coastal pool replete with red-legged frogs and their jelly-blob eggs, then make two last strategic stops. Checking under boards at an abandoned sawmill usually turned up at least one rubber boa, temperate outlier of a mainly tropical lineage, and in a roadside gorge further inland, an especially gnarly old log sheltered a Pacific giant salamander with the heft of a zucchini, its marbled copper and brown pattern recognizable from previous years. In two days we'd seen twenty species of herps, and Monday's lecture felt more like a gathering of friendly conspirators than the usual classroom scene. Word on the street was that our second trip would be even better, the *pièce de résistance* of the course.

In early May we'd load up on a Thursday afternoon and caravan south through the Central Valley, over Tehachapi Pass and into the Mohave. My plan was to stop after dark on a sandy two-track, miles from the nearest paved road and just outside the Bureau of Land Management's Desert Tortoise Natural Area; my troops would thus wake up without a clue as to what awaited them, smelling creosote bush and listening to horned larks. The sun's heat prevented remaining in the sack much past dawn, and answering nature's call revealed flowers everywhere, along with abundant flies and other insects that lizards eat. After yogurt and cereal I hustled everyone to pack up and hunt herps, but first we assembled for a minilecture about tying dental floss lizard nooses on fishing poles. I also instructed students on grasping lizards by a thigh instead of the abdomen, so the little reptiles neither struggled free nor were injured by squeezing.

That first morning was spent walking through open scrub. Students discovered tiger whiptails jerkily foraging among the scruffy plants and were astonished as zebra-tailed lizards zoomed off on hind legs—"like miniature dinosaurs!"—then paused, wagging their namesakes. Inevitably someone expressed delight at the soft, spiny skin of a desert horned lizard, picked up while it fed on an ant column or basked atop a sandy road berm. We snuck up on courting desert tortoises, once sitting just yards from a mating pair, the male grunting with each rhythmic thrust as grass-green liquid dribbled from his nostrils. We puzzled over a long-nosed leopard lizard's bulging sides, realizing she wasn't gravid when gaping jaws disclosed the hind end of a whiptail in her throat. And often as not, strolling toward the vans for sandwiches and juice, we were electrified by the defensive outburst of a yard-long Mohave rattlesnake.

Following lunch we drove a hundred miles east to Pisgah Lava Flow. My first few years at Berkeley, the class camped there, enduring howling windstorms and grit-laced pancakes; thereafter a stopover still allowed us to match wits with herps on a volcanic substrate. Within minutes students spotted little sooty-black side-blotched lizards, whereas foot-long Great Basin collared lizards are so well camouflaged and wary that, despite thorough scanning with binoculars for telltale heads on the lava spires, they were rarely detected. Long-tailed brush lizards look like creosote bush bark; although Stebbins's *Field Guide* advised slowly searching individual branches,[5] we managed to locate these late-afternoon sunbathers by high-grading for well-lit, leafless patches on the west side of bushes. Each trip somebody spooked a chuckwalla; rather than destroy its crevice lair with a crowbar, I'd rely on another Stebbins trick to examine it more closely: drape a dark shirt over the crack, tickle from below with a twig—"like a snake's tongue," as Bob told me—and when the big lizard moves up into fake safety, grab it!

Friday through Sunday my gaggle of naturalists camped in the Granite Mountains, fifty miles northeast of Pisgah. We caught desert night lizards under Joshua tree logs, afterward carefully replacing the tiny reptiles and their shelters. Peering under boulders the size of cars, we discovered southwestern speckled rattlesnakes, including a two-pounder that had eaten a desert cottontail. One morning as I savored scrambled eggs with *salsa fresca,* a student ambled up, said she had "something nice," and pulled a supple rosy boa out of her sweatshirt pocket—prompting discussion of its vestigial, clawlike hind legs and the biogeography of its giant kin. At Kelso Dunes we noticed how the Persian rug hues of Mohave fringe-toed lizards match red and brown sand grains; we tracked sidewinders to rodent burrows, then found them out crawling after dark and contemplated the scene by moonlight. Later we cruised nearby roads for rattlesnakes, less frequently finding a Baja California lyresnake or other prize.

For twenty years those courses were personal mileposts as I watched wild places change lives. My students produced t-shirts declaring "Largest Herpetology Class in the History of Western Civilization" and read *50 Simple Things You Can Do to Save the Earth.*[6] I was mooned countless times, and we ate well. There were the occasional sunburns and briefly lost undergrads; a woman dancing by the Eel broke an ankle, and a three-car pileup near Kelso accentuated my medic-inspired paranoia for safety (fortunately, no one was hurt badly). More memorably, though, by seeing forty species of herps in places as different as redwood groves and dry lake beds, we thought more deeply about organisms *in* nature. Back in 1979, a letter to Bill Pyburn began, "Survived a year in the California fast lane, unjaded if not unscarred. If there were ever doubts about preferring deserts this spring's teaching and fieldwork convinced me."

First, though, I had to persuade the senior professors to let me join their ranks.

After six years most universities either promote assistant professors to associate with tenure or fire them, and Berkeley's expectations were as high as the stakes were obvious. At a luncheon for new arrivals the vice chancellor cheerfully advised us, "Three things count here—research, research, and research."

That same semester M.V.Z. mammalogist Bill Lidicker wryly noted that my job entailed "fifty percent curating, fifty percent teaching, and fifty percent research," and Bob Stebbins warned against bogging down in education and conservation or soon I'd be doing neither. Promotion clearly would depend on teaching well, getting grants, publishing in prestigious journals, and garnering favorable evaluations from scientists all over the world. So the nagging worry was, what to study next?

My dissertation distinguished ancestral shared traits, like serpentine constriction, from those evolved independently, such as threat displays, and now I envisioned other implications of that classical comparative approach. Darwin and Wallace invoked selection *among individuals in populations,* whereas their theory sought to explain historically fixed adaptations, like limbs and fangs, of *entire taxonomic groups*—but forty years after Henry Fitch linked gartersnake diets with anatomy,[7] evolutionists still hadn't explored how to connect advantageous traits in modern populations to ancient legacies. Ecologists also ignored history while attributing species diversity differences to resource availability and competition, despite, for instance, the obvious effects of long-standing isolation on Australia's marsupial fauna. In short, "descent with modification" had inspired my graduate work, and now I struggled with new questions in that vein. Lizards helped sort it all out, even as my interests turned to their most successful limbless offshoot, the snakes.

That first spring Stebbins had taken me to his favorite teaching site, one I soon treasured too. Rocky terrain predominates along Mohave slopes and washes, but at Pisgah it spreads from a four-hundred-foot-tall volcano as ninety square miles of orange-smudged black moonscape. Out on the flow, low ridges, mineshaft entrances, and other refuges from the shimmering heat rarely exceed shoulder height, as if we are lumbering giants; creosote bushes encroach from the surrounding flats, lending flowers and fragrance to a withering cinderland. Although no dunes interrupt the rubble, sand accumulates within shallow interior pockets, and in them grass blades droop and twirl like drafting compasses, their circular tracks enhanced by afternoon shadows. Within an hour of our arrival I'd glimpsed melanistic creatures scuttling over the rocks, as if they too had burst from some fiery cauldron.

Well-camouflaged reptiles had initially attracted Stebbins, a gifted artist as well as herpetologist, to Pisgah. In this environment, conspicuous animals are more vulnerable to loggerhead shrikes and other predators. Thus black chuckwallas and side-blotched lizards blend into monochrome territories, the dark-hued flow, while black desert horned lizards look like lava stones in the sandy patches through which they forage. Conversely, tan side-blotched and desert horned lizards inhabit the surrounding scrub and hardpan where darker ones would be obvious, selection presumably favoring each version in its respective homeland. The foot-long western shovel-nosed snakes, however, vary kaleidoscopically, encompassing morphs that elsewhere typify entire regions—gray, salmon, or yellow with black bands, sometimes orange- or black-blotched within the light hues. Even such small snakes wander more widely than many lizards, and perhaps here they traverse so many backgrounds that no one version can prevail.

Diversity is higher in this volcanic Shangri-La than elsewhere in California because typically juxtaposed habitats intermingle, so Pisgah also fueled my interest in ecological communities. Side-blotched lizards, Great Basin collared lizards, and chuckwallas defend rocky redoubts while desert iguanas, zebratails, long-nosed leopard lizards, desert horned lizards, and tiger whiptails patrol the flats. Long-tailed brush lizards favor shrubs and Mohave fringe-toed lizards own the sandy pockets, coexisting such that, along with secretive western banded geckos and desert night lizards, a dozen species might compete for food and other resources. Southwestern speckled rattlesnakes and Baja California lyresnakes prowl ledges above sand-loving western shovel-nosed snakes, spotted leaf-nosed snakes, glossy snakes, and sidewinders, each of them encountering generalist western threadsnakes, desert nightsnakes, gopher snakes, long-nosed snakes, and coachwhips. I never figured out if the occasional desert tortoise was scratching out a living on the flow or just passing through.

Natural selection was evidently powerful under such harsh circumstances, and two Berkeley grad students pursued that topic, inspired by our trips to Pisgah and Stebbins's favorite reptile. Claudia Luke examined the natural history of fringe-toed lizards globally and showed that their foot specializations arose twenty-six times, in predictable relationships with habitats.[8] Sand dwellers in Old and New World deserts—iguanas, geckos, skinks, and so forth—have triangular fringes, whereas water-running Neotropical basilisks and Asian water dragons independently evolved squarish, flaplike projections. Meanwhile, John Carothers mimicked a variable population of Mohave fringe-toed lizards by clipping the namesake scales off some individuals, then used a miniature racetrack to prove that on sand they ran slower than unaltered animals.[9] Although Claudia and John hadn't demonstrated ancient differential reproduction—the actual process of selection—they'd made a strong indirect case for fringed toes as locomotor adaptations.

I soon pondered these issues on a different continent thanks to another Ph.D. student, whom I coadvised with Rob Colwell. Fabián Jaksic's data from Chile yielded surprising insights when analyzed with Fitch's Sierra Nevada findings and research on other "Mediterranean-type" ecosystems. We showed that only diet items identified more precisely than "beetle" or "rodent" portrayed whether predators truly ate the same prey—important because competition only occurs when species rely on the same scarce resources.[10] Moreover, ecologists typically investigated "feeding guilds" of, say, hawks and owls or mammalian carnivores, but three Chilean raptors overlapped in diet with a fox more than with other birds—so guilds would be better defined by specific prey than by predator taxonomy. Most intriguingly, reptiles at Fabián's study sites included only two snakes, a handful of small iguana kin called tropidurines, and a stout-bodied relative of North American whiptails, yet for so few species they were diverse in diet and habitat.[11]

Having experienced Pisgah's riches firsthand, I was keen to see simpler arid South American ecosystems. I also was fascinated by carnivory from reading about Komodo monitors and other Old World varanids, as well as from watching long-nosed leopard lizards consume other species half their size, and I knew the Chilean whiptail's only close relative to be the yard-long monitor tegu of Peru—where locals rounded up a smaller seed-eating Sechura Desert whiptail with portable corrals and roasted them for fiestas, complaining that monitor tegus ate the herbivorous species.[12] In short order I examined stomachs of the few museum specimens, finding arthropods, lizards, and a parrot's foot, perhaps taken as carrion. Monitor tegus, though kin to our whiptails and racerunners, *looked* like Old World varanids, so in the fall of 1980 undergrad Angus Wynn and I spent a month in Peru studying this giant lizard, followed by two weeks in Chile with Fabián.

Despite minor fiascos and near-catastrophes, our trip was worthwhile—the more so as an escape from the strain of my second divorce soon after moving to Berkeley. While our requests for collecting permits were being processed in Lima, Angus and I took a train ride to Machu Picchu, its ruins so impressive I vowed to someday walk the Inca Trail and view them from above. Caught up in Pablo Neruda's poetry and Gabriel García Márquez's *One Hundred Years of Solitude*,[13] I daydreamed of dying by plane crash or firing squad. Then a few nights later, finally under way, our jet shuddered during takeoff; minutes later

Angus pointed out at an engine spewing flames, whereupon the old man next to me gravely crossed himself and said, "Señor, la máquina está mala." I stared into darkness as the plane banked a steep descending turn, recalled hugging my parents, and wondered if I might actually see the fireball. Instead, we returned to Lima without incident, were herded onto another plane without explanation, and reached the northern town of Chiclayo after a dreamily smooth flight, over moonlit clouds that looked to me like perfectly spaced, snow-white acacias.

That evening we paid too much for a cab to the car rental agency, thanks to my rattled psyche and clumsy Spanish (Angus hadn't even learned enough to read traffic signs). The next morning we found our trunk popped and much of our field gear stolen. After breakfast we headed north anyway, into an alien landscape. Sandwiched between mountains and ocean, one of Earth's driest places because of the offshore Humboldt Current, the Sechura Desert dominates northwestern Peru. Chiclayo, on its southern edge, and Piura, at its center, though sizable cities, are surrounded by glaring barren flats or sandy plains populated by terrestrial bromeliads and mesquite, occasionally interrupted by rocky *cerros* and farming villages. We jokingly diagnosed the scrawny cattle as alive or dead based on verticality, then discovered that one cow sheltered three species of lizards, her withered legs serving as handles when we checked the carcass. And on a highway stretch that threatened to strand us without gas, as if *la máquina mala* hadn't quashed my morbid fantasies, Angus hurtled us through a checkpoint in which the soldier slept propped on his rifle. "Holy shit," I blurted out, visualizing the Bolivian fusillade that cut down Paul Newman and Robert Redford in *Butch Cassidy and the Sundance Kid,* "an octagonal red *Alto* means stop!"

Andean condors soared over the sparkling Pacific seashore on the last day of our month-long circuit, but we saw little other wildlife except small birds and seedpod-eating foxes. Although the Sechura reptiles comprise several geckos and tropidurines, the whiptail relatives, two racerlike serpents, and one each worm-lizard, coralsnake, and pitviper, we searched in vain for snakes and encountered only three of our elusive giant lizard quarry. The largest monitor tegu sped across a road near Olmos (in hindsight, I realized I should have recruited local children to search for it), and we watched another prowl among boulders and columnar cacti, looking like a miniature version of Komodo monitors I'd seen on TV. The one we caught, a handsome black male with yellow markings found near Talara, on the continent's western-most tip, was finally dug out from a yard-deep, eight-foot tunnel that also contained shed skins of worm-lizards.

We didn't learn much about the big lizards, beyond admiring their long forked tongues and pebbled skin, their jagged teeth and razor claws, but those weeks in Peru enhanced my curiosity about adaptation. If examining preserved *Tropidurus thoracicus* was thought-provoking,[14] noosing those wriggling sand-dwellers, so much like the distantly related but ecologically similar Mohave fringe-toed lizard, was downright inspiring. Holding a live monitor tegu was off the charts! And beyond such impressive convergent evolution, we saw never-before described behaviors. On the coast, ten-inch *T. peruvianus* dashed from rocks above the tide-line down onto wet sand, scarfed up arthropods, and fled back from the crashing surf; a few miles inland, dozens of herbivorous Sechura whiptails gathered in the shade of an algarrobo tree, striped skins awash in aquamarine hues, and vigorously hand-waved at each other—alas, to no effect, as far as I could tell.

Soon enough we were gazing on snow-covered peaks of the continental spine glinting off to the left throughout our four-hour flight south to Santiago. Over the coming days Fabián would introduce us to his country, Nobel laureate Pablo Neruda's homeland, but I couldn't shake the more general problem of specialization and kept daydreaming about how all North American and South American fringe-toed lizards possess adaptively modified feet, implying that their respective ancestors did as well. Here was the problem: Selection obviously can shape population genetics, but how might we test whether that process, deep in geological time and coupled with environmental opportunity—in this case, running on sand—had favored now *invariant* traits in particular lineages? Two other troublesome examples further stirred my thinking.[15]

In the 1960s, University of Texas professor Eric Pianka initiated research on desert reptiles, including some textbook examples of convergence. Our dozen-plus species of horned lizards, related to other New World iguanians, are slow, tanklike, and superbly camouflaged; they prey almost entirely on ants. Eric's studies also revealed that the squat-bodied, spiny Australian moloch, kin to bearded dragons and other Old World agamids, is comparably specialized: one stomach that he inventoried contained a mind-boggling twenty-five hundred ants! Those features that independently evolved in horned lizards and moloch he attributed to selection imposed by tiny, toxic, chitinous prey, so low in quality that enormous numbers must be consumed, leading to huge stomachs that preclude speedy escape from danger. On the ground, though, I couldn't help wondering about a Chilean tropidurine, *Liolaemus monticola*, that eats only ants, hides among rocks, and looks like nothing so much as western fence lizards in California chaparral. If a narrow diet imposes such all-powerful selection, where were its horns?

Monitors were perplexing too. At the time, biologists explained their sharp teeth, flexible skulls, and unusually high physical stamina as specializations for eating large vertebrates, without, as it turned out, much in the way of supporting natural history data. Yet when my grad student Jonathan Losos and I examined museum specimens of varanids for comparison with long-nosed leopard lizards and monitor tegus, we were taken aback as one after another good-sized individual contained beetles, snails, and other diminutive prey. A two-pound, yard-long Bengal monitor proved typical of thirty species we studied, having eaten some two dozen insects and a lizard—its heaviest item, and the only vertebrate prey, weighing less than an ounce. A few varanids do specialize on mammals, while at least one eats mainly fruit, but they're deep within a family tree of generally large lizards that we proved consume mostly small items.[16]

Conventional wisdom about monitors amounted to just the sort of "adaptive storytelling" ridiculed in a 1979 paper by two Harvard biologists,[17] Stephen Jay Gould, who had achieved fame for his masterful *Ontogeny and Phylogeny*[18] and monthly *Natural History* magazine essays, and Richard Lewontin, highly regarded as a molecular evolutionist and leftist intellectual. Together they'd vehemently opposed colleague Ed Wilson's *Sociobiology*,[19] complaining that his enthusiasm for genetic determinism and selectionist explanations of human behavior encouraged racist politics. Now they made sweeping arguments that adaptation is difficult to prove and that the side effects of design requirements or, in humans, culture play more important roles. Sure, they admitted, selection might account for moth wing colors, but not skull differences between horses and lions, let alone anything involving our behavioral norms. The ensuing squabbles were also prone to authoritarian rhetoric—Harvard's Ernst Mayr asserted that convergent evolution provides ample evidence for ancestral selection[20]—but they fostered a more widespread, healthy skepticism.

Henry Fitch might have smiled at Mayr's failure to link natural history more directly with evolutionary history, given the elder biologist's earlier criticism of his gartersnake dissertation, but Steve Gould's role in rehabilitating "adaptation" was far more ironic and influential. However brilliant, here was an unabashed urbanite who scoffed at affection for warm and fuzzy creatures,[21] a polymath who studied fossil snails, downplayed predation and other selective factors, and claimed that exotic organisms are superiorly adapted—"astonishingly," as mammalogist Tim Flannery noted, since invaders lack natural predators and thus "this argument takes no account of ecology."[22] Nonetheless, Steve's 1982 proposal with Elizabeth Vrba that preexisting traits are *exapted* for novel uses supplied a pivotal insight.[23] As Losos and I soon demonstrated with monitors, evolutionary shifts in ecology and performance, exaptive or adaptive, are detectable using the same rationale with which we infer changes in anatomy.

Simply put, attributes of diverging lineages that are lacking in close relatives probably arose in their common ancestor—the upright walking and savanna-dwelling lifestyles, for example, characteristic of australopithecines and us but not of chimps, bonobos, and gorillas. Whether such comparisons prove that ancient selection fixed those traits in a population seems largely a matter of taste; for many evolutionists that's the only plausible conclusion, whereas others, myself included, prefer sticking to what we can infer more directly.[24] In any case, convergent similarities between horned lizards and moloch *are* correlated with feeding in open areas, and the rock-loving Chilean ant specialist can't refute the hypothesis that the others' spines *are* antipredator adaptations. Komodo monitors do take deer and pigs, but ancestral varanids ate small items. So perhaps their raptorial jaws and mammal-like endurance originated for wide-foraging and rapid capture, regardless of prey size.[25]

My data from field observations and museum specimens addressed interesting questions. The probationary years at Berkeley zoomed by, and, thanks to supportive colleagues there and elsewhere, I cleared academia's "publish or perish" hurdle. Studying desert lizards led to research on snake ecology, venoms, and parental care, topics that still captivate me; arid lands also afforded tranquillity and a sense of going wild, even though I never lacked water or traveled far on foot—only years later would I read horrific accounts of dying of thirst in Mexico's Gran Desierto and, while backpacking in the Barranca del Cobre, come to more viscerally appreciate my own adaptive shortcomings. In the meantime, there were plenty of ups and downs outside of science, and where better to ponder them?

In May of 1985, freshly tenured and preoccupied by my dad's sudden passing, I was helping Claudia Luke with her doctoral research at Pisgah. We were walking along chatting, hot and hungry from a morning of chasing lizards, when only a dozen yards ahead an enormous magenta coachwhip materialized on the gravel track. Named for the braided look of their scales, with birdlike eyes, these lightning-fast creatures occur in habitats as diverse as pinewoods and desert scrub. More than most other serpents, they defy control—I've ripped the knees out of jeans diving for one on a paved road, been stabbed by a yucca while grabbing for another. In hand, they're prone to thrash wildly, strike at one's face, and if all else fails go limp and feign death.

This coachwhip looked like a piece of dazzling pink Hula-Hoop, and capturing it was instantly paramount. There was a split second of evaluation, as if we doubted the creature's existence, followed by an explosion of action; everything was over in seconds. We stumbled and scrambled while the coachwhip seemed to fly, not always touching earth. As I charged full tilt, the snake shot down a sandy embankment, then up an escarpment. Claudia cut diagonally toward the high rocks, trying to prevent its escape, but my feet slipped backward and I pitched through a somersault. Luckily my head was cushioned by a creosote bush at mid-roll as arms and legs tumbled over lava chunks.

I lay as I landed, propped on one elbow, and in an affectedly calm voice asked, "Did you get that snake?" Claudia shot me a skeptical grin as I passed from clarity to confusion, from relict obsession with the coachwhip to startling pain. Really addled now, my brain felt loosened from its moorings. I ached all over. Arms, legs, and clothing were shredded, my watch torn off. I was bleeding freely, matted with sand and bits of vegetation. Meanwhile the snake had sped into a crevice, so we retrieved my crumpled hat, then limped back to camp to pick out the grit and patch me up.

After an hour's absence we saw the coachwhip's head extended several inches up from the rocks. While I watched with binoculars and distracted our quarry with hand waving, Claudia lowered a noose over its head from behind, only to have the snake break the flimsy dental floss loop and disappear into its hole. Returning to camp for shade and water, grimly resolute, I made a new noose out of stronger material. We waited another hour, then crept over the lava and repeated the entire operation, but the snake slipped out when the thick cord failed to close tightly. Game for a third try, Claudia wove a stout pliable snare out of three strands of floss, and although she succeeded in approaching again, this time the snake dodged our trap. We conceded defeat, and I returned to M.V.Z., disgruntled and nursing various wounds.

Two weeks later Claudia brought in a brick red Pisgah coachwhip and, although I'd collected many reptiles for teaching and research, this snake made me uncomfortable. The caged animal was subdued, a victim of trophy-ism on my part, its colors not magical—the one that got away was surely bigger, brighter, faster. Obviously something besides education and science had motivated our breakneck pursuit, such that this second snake's capture felt unjustified, and I asked her to release it back at the lava flow. We'd been more predators than scholars that morning, engaged in nature's rough and tumble chase, and although brains and legs didn't carry the day, I'd gained an appreciation for serpentine locomotion beyond anything written in books.

Soon after that episode my Berkeley girlfriend and I split up, prompting a surreal six-thousand-mile odyssey to check out new places and see old friends. On a Mohave back road, emotional lid clattering like a cheap teakettle, I sang along with Bruce Springsteen's *Born in the U.S.A.*, flooring the little Honda through gully crossings and imagining myself airborne. When an elderly lady asked for help at a highway rest stop, I held her trembling husband up to a urinal, one arm around his chest and the other bracing against the wall while tattooed motorcycle thugs wisecracked all around us. By day Emmylou Harris's *Ballad of Sally Rose* blared from the tape deck as Joshua trees gave way to Sonoran Desert saguaros. Nights I endured spasms of guilt reading Marilyn French's *The Women's Room* because, although never abusive in the manner of her scathingly portrayed male characters, I was too often clueless and selfish.[26]

I reached the Chiricahuas during a late-afternoon thunderstorm, wolfed down a burrito at the Portal Store, and within minutes of driving up Cave Creek Canyon was astonished by a four-foot green ratsnake stretched out on the road. This semiarboreal species ranges from Costa Rican tropical dry for-ests to the Sierra Madrean sky islands of Arizona and New Mexico, but is nowhere common. Recalling that Albert and Anna Wright, authors of *Handbook of Snakes,* never found one in all their travels, I admired the ani-mal's gently tapered body, the long head with squarish snout and protuberant eyes, and I lingered over its subtle hues. Its English name notwithstanding, the Wrights referred to one sent them from the Pajaritos as "buffy or dull citrine,"[27] and mine was an olive blue-gray, reminiscent of agave leaves. A dec-ade of visits to these mountains would pass before another enigmatic green serpent crossed my path. More fortunately, on one of those sojourns I fell for Kelly, a charmingly intrepid desert lizard biologist with whom I've since shared the Inca Trail and so many other adventures (our move to the Northeast in 1999 ensued when Cornell offered her, then me, professorships).

That night, however, I drove back through Portal in solitude, brooding about snakes and women, then snapped to my senses dodging Mexican spadefoot toads that hopped across the rain-drenched pavement. After only a few minutes, barely into the San Simon Valley's mesquite grassland, I recognized a wriggling serpent in my headlights and slammed on the brakes, excitement rising as I scrambled out with flashlight and tongs. The Sonoran coralsnake, hiding its head under writhing red-, cream-, and black-ringed coils, waved a tightly curled tail, made hilarious popping sounds with its vent, and bit my forceps. Then for a grand finale the squirming sixteen-inch cobra cousin regurgitated a western threadsnake, as if to say, "You ain't seen nothin' yet!"[28] Later I fell asleep pondering whether the coralsnake gave up its meal so as to crawl away unburdened or if vomiting was part of a bizarre, multimedia strategy for repulsing predators.

My first trip to the Chiricahuas led to studies of black-tailed rattlers, but in 1985, with my fortieth birthday at hand, it amounted to an unexpectedly internal journey. I searched the slopes for rock and twin-spotted rattlesnakes, encountered them in lively abundance, yet couldn't escape mortality. One afternoon as I clambered over talus seeking herps, tortured moans drew me to a half-ton Hereford bull that had slipped off a cliff, hung by his horns between limbs of an oak, and died while I stood back from the flailing hooves. A week later I drove up through New Mexico to be in Rob Colwell's Colorado wedding and visited my paleontologist friend Len Radinsky in Utah just days before cancer took him, two years shy of fifty. Then I returned to Tucson for one last bout of snake hunting, highlighted by a lavender-and-charcoal banded tiger rattler found crawling among the saguaros. Another eighteen hours of driving, I was home in Berkeley, sad but lighter.

My desert travels haven't so much presented crisp epiphanies as an awakening that began in grad school and first came into focus during my wanderings that summer. No matter how glorious a Sonoran sunrise nor how mystical a Mohave moon, there are no natural laws of fairness and mercy, only lives unfolding according to heritage, chance, and the extent that we rise above them. Reptiles adapt to harsh environments or they don't persist; we wear the pain of death with varying discomfort. And as ever-optimistic Ed Wilson said in closing *Sociobiology,* we've got some time left—by which he meant for solving environmental problems but also to figure out our own existence.

There'd been Marsha's murder and the toddler I'd failed to save; now the twin challenges of grief and hope were manifest again with the passing of my dad and kind, generous Len. There would be more losses too, but also abundant joy and a measure of clarity in the everyday workings of nature.

For millennia arid lands have inspired humans to look beyond immediate concerns, seeking wisdom in stone tablets and the solace of wide-open spaces. That summer of 1985, though, I was awash in reckless self-pity, grateful for distractions and reminders that we mustn't despair. "Attend to matters at hand," the rattlesnakes and spiny plants seemed to say, whereupon small black lizards, having survived another day, might chime in, "Embrace it all, never surrender!" Months later, while I paused for coffee during a Bay Area drizzle, desolate places flooded my thoughts. I hoped the Pisgah coachwhip really was fluorescent pink and seven feet long, still reigning over the wildest parts of the lava flow. I recalled hunting fossils with Len in Wyoming's Bridger Basin, his disdain for academic status and pleasure at spotting Eocene mammal teeth in a sea of multicolored pebbles. I remembered, too, that my father's favorite pastime was photographing rocky canyons, and wished I'd asked him why.

EIGHT

Hot Wet Places

RAINFORESTS ARE DIMLY LIT and exceptionally diverse—claustrophobically dark and fecund—so no wonder tropical biologists end up puzzling over existential questions. At La Selva Biological Station in Costa Rica, giant trees with buttressed trunks tower overhead, obscuring the sky, and every glimpse holds the vibrant greens and somber browns of plants and their decaying remnants. After a torrential shower the air reverberates with the buzzes, whines, and clicks of insects. Mantled howler monkeys sound off in the distance. All around us leaf litter reeks from the chemical adventures of microbes, and over the course of hours my puny primate nose wrinkles toward some collared peccaries, then heaps of rotting fruit and a pile of cat droppings. Rounding a trail curve I'm baffled by a shimmering lavender stripe, dozens of yards long and a half-inch tall; then I drop to my knees and contemplate thousands of leaf-cutter ants, each carrying a single delicate flower petal. And from time to time, slogging along the muddy paths, I imagine being overgrown by mosses and fungi, or devoured by spike-headed katydids the size of small mice.

Setting aside matters of life and death for the moment, what do ecologists mean by "exceptionally diverse," and why might anyone care? A comparison among some familiar places illustrates how numbers of species increase toward the Equator, culminating in unparalleled tropical richness. California reaches from Death Valley's floor to Mount Whitney's summit, spans parched salt flats to drenched redwood groves, and yet across ten degrees of latitude boasts only thirty-five species of snakes. Almost twice that number occur in La Selva's five square miles, as if a house full of serpents were packed into a thimble, and there are nearly four hundred species of birds, more than half as many as in the continental United States. Tropical faunas encompass more lifestyles too, thanks to rampant adaptive diversification; most temperate bats feed on insects, for example, whereas some of their hothouse relatives specialize on fruit, nectar, fish, frogs, or birds.

This dramatic global variation has long intrigued naturalists, and its causes are partly understood. Rainforests usually occupy middle latitudes, so Earth's most biologically opulent regions are hot and wet. Some of them have been that way for millions of years, during which rising seas and tectonic events fragmented landscapes, catalyzing the origin of new species. More land, more sun, and more rain, coupled with geographical isolation and geological time, have fostered plant evolution—and thereby more plant-eating insects, insect-eating frogs, frog-eating snakes, and birds and mammals that eat them all. At local scales earthquakes, volcanoes, and windstorms annihilate chunks of habitat, which are then colonized by species that live in the resulting light gaps. Those sun-loving newcomers are eventually replaced by shade-tolerant species, so that natural disturbances further increase diversity by generating patchworks of succession in what at first glance appears to be unbroken forest.

Tropical biotas are also among the most endangered anywhere, their most charismatic inhabitants often difficult to find. Ecotourists adore emerald-and-red quetzals and iridescent blue morphos, and with coaching they might tolerate the jararaca pitvipers whose venom chemistry inspired a popular blood pressure drug. Predators are usually tough to see, though. Whereas in an hour a person might find dozens of snakes on a Missouri hillside, I averaged one a day at La Selva, and after twelve months of fieldwork I still hadn't seen all the species at that serpent-rich locale. Because rainforests don't offer Serengeti-like vistas, we can't drive folks through them in a safari van, striped like a baby tapir instead of a zebra, to show off the big cats. Instead, advocates need to cultivate perspectives that make those places *feel* wild, even if one doesn't see much that day. We should teach neophytes to flare their nostrils at unfamiliar odors, differentiate splayed tracks of jaguars from parallel-sided prints of mountain lions, and distinguish among the sounds of frogs and birds. With luck, visitors might overtake a white hawk, as I once did, so close on an overhanging limb that the immaculate bird seemed at first illusory.

My fascination with steamy venues began on a childhood sojourn in the Philippines, enhanced by reading Raymond Ditmars's melodramatic tales of bushmasters and vampire bats.[1] Years later, as a soldier I requested assignment to Panama, hoping to find exotic creatures and avoid combat, but was stationed in Germany instead. So my first tastes of the tropics came on grad school trips to Mexico and Guatemala. I've since enjoyed a decade of visits to La Selva and sporadic stints elsewhere in Latin America, Africa, and Asia. Studying feeding and defense in snakes justifies my travels, with diversity an overriding concern: How can so many species fit into hot wet places? Along the way I've also been enchanted with other predators, as well as impressed by how local peoples' lives play out and dismayed by the loss of tropical habitats. Just as deserts afford simplicity and clarity, I've learned, rainforests exemplify complexity and obscurity.

Snakes are notoriously difficult to find in rainforest, whether rare, hard to see, or both. In the 1960s Henry Fitch had heard that La Selva was unusually snaky, and after a brief reconnaissance he planned several Costa Rican research transects. Former student Donald Clark was set to serve as postdoctoral fellow, but Don got a job at the last minute and his replacement, a mammalogist, mostly caught bats instead of snakes. Henry ended up studying lizards elsewhere, but he told me about the field station, and, encouraged by Rob Colwell, my Berkeley colleague who'd worked there for years, I headed south in search of giant vipers.

About the size of Virginia or Austria and proudly committed to conservation, Costa Rica encompasses an impressive array of habitats.[2] Soaring cordilleras accentuate the country's long axis, separating Atlantic and Pacific coastal plains, and its capital is on a central plateau. Getting to the Caribbean versant in 1981 was an adventure. I flew in a small plane between volcanic peaks that rise to more than eight thousand feet around San José, with a little girl in front of me screaming throughout the stomach-churning ride. There followed a dirt-strip landing, billowing red dust and flapping banana leaves, and an hour's drive on a horrible road to La Selva. For the next few years I traveled back and forth to the field station in *El Rápido,* a dilapidated red bus that required nine hours to traverse the pot-holed highway; by 1991, however, we'd get there in an hour, winding down a freeway through the sparkling mountain scenery of Braulio Carrillo National Park.

Travelers often speak as if "rainforest" and "jungle" were synonymous, whereas biologists restrict the latter to regenerated vegetation after forest has been cleared, either naturally, as with a wind-toppled tree, or by humans. Also known as second growth, jungle is of short stature, with a dense understory and difficult to walk through without cutting trail. Mature rainforests are multilayered, reaching, in the case of the La Selva canopy, upward of 160 feet.

Shorter trees shade a fairly open understory, within which trails are constructed more for navigation than access. Woody vines and epiphytes are obvious but not abundant; palms are common and diverse, encompassing emerald-leaved *Geonoma* the size of small shrubs as well as big, spiny-trunked and stilt-rooted *Iriartea*. Only about 1 percent of sunlight reaches the forest interior, as irregular solar flecks and blotches, with temperatures generally mild by tropical standards.

La Selva appealed to me for its mature rainforest, almost sixty species of snakes, and other researchers who might help me find vipers. I especially wanted to study bushmasters and terciopelos (often inappropriately known as fer-de-lance), among the largest venomous serpents, and rumor had it they were common. Soon after my arrival, though, I found carnivore droppings and hatched a plan to assay all predator diets, paralleling the Chilean studies with Rob's and my student Fabián Jaksic but in a much richer tropical assemblage. Typically I'd set out after breakfast, sometimes accompanied by an assistant and others who hoped to see snakes. We wore light pants and long-sleeved shirts, rubber boots, and bandana sweatbands. My pack bulged with camera and telemetry gear, other research gadgets, and water bottles. We'd walk all morning, paying special attention to long dark objects, because sometimes instead of roots that might trip us they were snakes that could bite. For weeks on end, during some dozen trips, I recorded observations, collected scats, and marveled at my good fortune.

Arthropods, birds, and mammals are the most obvious animals at La Selva, with many insects distinguished by dazzling colors, nasty stingers, or both. Inch-long, shiny black *bala* ("bullet") ants abound, and wasp nests seem preferentially hidden under leaves at face height. During our walk an electric whine crescendos, then ebbs as a cicada chorus goes briefly quiet. The Costa Rican avifauna encompasses plenty of small drab birds but also gaudy toucans and plump tinamous, their haunting calls like flute music at dusk. We'd occasionally get a glimpse of collared peccaries and rarely find the dinosaurlike tracks of Baird's tapir. Among otherwise ratlike rodents are pacas and agoutis the size of house cats, and who can deny the appeal of wild primates? From time to time white-faced capuchins chattered threateningly from nearby trees, while cartwheeling spider monkeys high overhead left our necks aching from watching them. And I was never too tired to enjoy the roaring howlers as darkness fell in the forest.

As for the surrounding greenery, certain easily identified species make exceptionally high plant diversity palpable even for those focused on vertebrates. Monkey pot trees, for example, have corrugated, straight-boled trunks and skull-sized seedpods that accumulate around them. I knew each of these majestic "canopy-emergents" along several miles of trails, like familiar commuters on a bus route—attentiveness no doubt inspired by fear of falling "monkey pots" and curiosity about why frogs don't use their rain-filled husks for breeding ponds, as happens with a related Brazilian species. Without consciously counting individuals of this species, I was impressed by their scarcity and realized, because I'd been told most tropical trees are likewise rare, that the total number of *kinds* in the forest must be great indeed.

Of course, species richness is most obvious for familiar organisms, and La Selva bustled with herpetological treasures. During a morning I'd see dozens of strawberry dart-poison frogs, and the leaf litter buzzed with their territorial calls. Occasionally a North American wood frog look-alike careened across the trail; with the jumper in hand, squirming red-and-yellow spotted legs confirmed it was *Rana warzewitschii*. I'd snatch little brown frogs, and each time careful inspection disclosed yet another species of *Eleuthero-dactylus*[3]—one with black mask, another with dark flank bars, while a third was warty and filled my palm. I'd dive on a shiny black snake basking in a sunspot, then palpate a juvenile smoky jungle frog and an adult narrow-mouthed toad from its stomach. Later, as we greedily drained canteens before heading back for lunch, someone would say with amazement that almost every frog we'd found was a different species.

Studying all of the roughly 115 species of vertebrate predators at La Selva proved impossible, but our findings, coupled with previous work, elucidated some details of their natural history and contradicted prevailing wisdom.[4] Ecologists at the time believed that more species coexisted in the tropics by having narrower, often unique niches—and indeed, ocelots do mainly eat rats, harpy eagles mostly sloths and monkeys, while vampire bats subsist entirely on blood, and some snakes specialize on slugs. We were surprised, though, by the number of dietary generalists. Eyelash palm-pitvipers prey on frogs, lizards, birds, mouse opossums, bats, and rodents, while crane-hawks methodically search nooks and crannies of tree trunks and epiphytic plants, taking everything from caterpillars and katydids to baby birds. Jaguars consume animals ranging in size from small lizards and armadillos to deer and tapirs.

As in the simpler Chilean predator assemblages, we also discovered unexpected diet overlap. Jaguars eat sloths, perhaps catching them thanks to the inverted creatures' puzzling habit of descending from the canopy to defecate, as well as green iguanas that come down to nest—and because brown vinesnakes and double-toothed kites prey upon the baby lizards as well, those two-hundred-pound cats might compete with a quarter-pound serpent and a small hawk for food. Each species thus faces diverse enemies in an intricate food web, enforcing a perilous uncertainty on rainforest lifestyles. Plants and animals have evolutionarily responded to these challenges with spectacular adaptations—some katydids, for example, have fake herbivore damage on their leaflike wings. In fact, our fascination with camouflage, chemical defenses, and other protective mechanisms traces to pioneering tropical biologists like Alfred Russel Wallace and his friend Henry Bates, who discovered mimicry while collecting Amazonian butterflies.

Annoying, even painful incidents inevitably punctuate fieldwork, reminding us not to take nature for granted. Each trip I slipped on a muddy trail, grasped wildly to break my fall, and learned anew that some plants come armed with shockingly effective spines. Fungi grew on camera lenses and beneath my beard, and a nematode under the skin of an index finger itched wickedly until I killed it with industrial freezing spray. Nonetheless, blissful times overshadowed unpleasant ones, as when our students rendezvoused in a swamp with University of Michigan's John Vandermeer and me during a solar eclipse. Insects and frogs called as the sky darkened, and we spontaneously whooped with delight at the incongruity of nocturnal songsters at noon. One feels a welcome distance from humanity in remote places, but during the eclipse everyone shared a cosmic moment, against which we were incomprehensibly small yet connected.

Life is all about water, and rainforests are blatantly dependent on moisture. We are a mile out on the central trail at La Selva when a storm roars in, and as two of us make for a thatched shelter, or *techo,* my Tico companion remarks that San Pedro must be moving furniture in heaven. The sky darkens dramatically, and first one, then several more lightning cracks segue into long rumbling thunder. Shortly thereafter a different sound sneaks into my consciousness, as billions of raindrops hit millions of leaves, quickly becoming a deafening orchestral downpour. Water falls around us in colossal sheets, as if a gigantic pot were being emptied directly overhead, and pours off the eaves in gloppy, twisting strands that splash my feet from a yard away. After ten minutes a stream two feet wide has replaced what had been a shallow trailside ditch. Already the rain is diminishing. Light comes back in sparkling golden sun flecks, and everywhere leaves and bark glisten. The air is redolent with forest aromas flushed out by water. Here this is all as normal and as epic as sunrise.

Water makes its way to the oceans, then evaporates and cycles back as condensation or precipitation. Streams, like storms, are obvious in tropical landscapes—witness the Earth's largest river system, straddling the Equator and draining the Amazon Basin. Kelly and I once took a ferry from Manaus across that gargantuan watercourse, watching from an upper-deck rail during the sunny two-hour voyage. Dolphins breached alongside us. About halfway across we hit the mixing of "white waters and black," made famous by Wallace, Bates, and other explorers. Leaving port, we'd chugged through the dark tea flow of the Rio Negro, coming from Venezuela to the north; then the caffe-latte Rio Solimões churned across our bow, its headwaters to the southwest in Peru. Now they roiled along in midstream, braided and discrete, as if elementally different liquids struggled with immiscibility.

Those rivers collectively drain the Guianan highlands and Andean slopes and then, as the Amazon, empty into the Atlantic Ocean a thousand miles after their source waters have mingled in central Brazil. On a much smaller Costa Rican scale, the Río Puerto Viejo originates on the flanks of Volcán Barva and picks up small tributaries as it flows in a broad arc, first east and then north-northwest onto the coastal plain. Once a pristine liquid groove in an unbroken expanse of forest, today it carries runoff from surrounding pasturelands, and these opaque waters join the clearer, straighter, and rockier Río Sarapiquí soon after passing La Selva on their way to the Caribbean. At the time of my first visit the river still formed the station's northern boundary, and everything arrived and left by it. Kitchen supplies came on a boat, as did staff, students, and researchers; one night six of us waded our groaning cook across on a litter when he was stricken with what we feared was a heart attack but proved to be only gall bladder trouble.

A placid stretch of the Puerto Viejo in front of the station was popular for swimming, and no doubt many a problem has been deciphered there over beer and a quiet sunset. Nonetheless, at times the river was rebellious, reminding us that life wasn't always tranquil. One afternoon it rose so fast from storms upslope that I measured an inch every five minutes as water lapped up the steps; watching its contradictory moods, I more easily understood how Latin American writers so persuasively blur what gringos distinguish as reality and magic. Huge trees roared by in the foamy brown current; on one of them was a dead pig, with skin discolored and parched ears erect, facing downstream as if on some hellish last journey. Then, in 1982, the station gained a touch of modernity with construction of a suspension footbridge, after which everyone walked back and forth to take meals in a shiny new *comedor* on the north bank. Some older hands resented these changes, but the flora and fauna probably benefited from concentrating our activities away from the forest.

Five years later, on sabbatical leave from Berkeley, I sought fresh perspectives from the bridge. Ringed kingfishers, keel-billed toucans, and other flying dinosaurs perched on the spits and tree limbs that extended from the downstream riverbanks. Surreptitious squirrel cuckoos with lemon bills and sorrel plumage scoured foliage for lizards and caterpillars, and a king vulture soared overhead, recognizable even to novice birders like me. The only prominent asymmetry was a fallen tree reaching from the south bank into midriver, where it snared logs and other debris. Green ibises perched there in misty dawn light, whacking their turquoise bills in social ritual; later on, spectacled caimans, close kin of our alligators, basked on the toasty sunlit lounge. One morning a Neotropical river otter lay on the logjam and groomed her svelte belly, swollen nipples visible, then rubbed her rump against a stump and stretched out for a nap. I stopped again after lunch and observed two of the willowy weasel relatives hunting around the submerged trunk, after which each emerged crunching a large crustacean.

The bridge also offered an excellent vantage for watching arboreal vertebrates, most of which are difficult to see and poorly known. Three-toed sloths were sometimes visible in nearby *Cecropia* trees, and one morning I photographed them waging "hand-to-hand" combat. A sloth was curled up asleep as I trudged over for breakfast, and a couple of hours later it was leisurely eating new, orange-brown leaves in the same tree. After eighteen minutes a second animal emerged from adjacent vegetation, ascended the *Cecropia* at twice normal sloth speed, and pummeled the resident with heavily clawed forefeet. Both were males, judging from their yellow and black dorsal markings, and after a surprisingly vigorous and noisy struggle, resident chased intruder from his tree at *three* times normal speed. Sloth sociality, my serendipitous observations implied, is more complex than previously recognized.[5]

Until that year my visits had been between February and August, when green iguanas are dull olive and well camouflaged. In late November, however, the sides and dorsal spines of males had turned an electric coppery-orange, and from the bridge I easily counted more than a dozen sunning along a quarter mile of riverbank. Basking was followed by head-bobs and dewlap shakes that established who was most attractive to females, and by noon their abdomens bulged with a morning's meal of leaves. One day Isaías Alvarado, station foreman and lifelong area resident, walked up as I was contemplating iguanas. "Those lizards are always easier to see during these weeks," he noted matter-of-factly, and then asked, "Do you know why?" I'd just been thinking about how Stan Rand and I had used collecting dates for baby iguanas in museums to demonstrate that breeding varies latitudinally; so at La Selva mating happens now, as the dry season begins, and eggs hatch months later, with the onset of heavy rains—coinciding with what Isaías had seen many times, a seasonal change in male colors![6]

Besides people, transients on the elevated corridor included prehensile-tailed tamandua anteaters, seven-foot-long tiger ratsnakes, and raccoons. A hog-nosed skunk occasionally waddled over to raid kitchen garbage, and I was reminded of Robin Hood's skirmish with Little John as researchers edged against the rails rather than turn back from the well-armed scavenger. And soon after María Marta Chavarría set up an ultraviolet moth trap over the river, a brick-and-cream-furred, lavender-scrotumed woolly opossum plundered her catch. We stared in awe as the gaudy bridge possum, most arboreal of La Selva's five species of marsupials, scampered across thirty-five-foot-high suspension cables instead of using the heavily traveled walkway.

High views from forest openings have intangible virtues as well. Out on the bridge over the Puerto Viejo, free of the all-encompassing botanical livery, one encountered a preternaturally elemental world. Dusk colors changed from green through purple and scarlet rather than simply fading into obscurity, and the multidimensional trees turned sharply flat before dissolving into blackness. On clear nights the stars and Jupiter evoked odd memories of cooler climates. Perhaps these impressions stemmed from the quality of light not yet filtered through canopy layers, reactions of the retina to more contrasting surroundings, and even fatigue, but I felt content to admit that perceptions of reality are far from absolute. I was also reminded that the most urgent challenge for conservation is for us to fit in with landscapes and not destroy them. The nicely weathering bridge, its green paint blending in with vegetation, symbolized that effort.

With all that Costa Rican experience under my belt, in November of 1990 I jumped at the chance to study an Old World tropical snake fauna. I'd spent nearly a month reconnoitering a montane African rainforest, and my face was haggard, clothes sweat-soaked and disheveled. Yet I was grinning because, despite exhaustion, I'd kept up with the shorthaired fellow in front of me. Vincent Bashekura was a Ugandan game guard holding a bolt-action .30–06 rifle and wearing khaki fatigues that looked freshly pressed. His demeanor reflected earnest pride in protecting the gorillas of what British colonists named the Impenetrable Forest, and my triumphant smile was illusory. In truth, Vincent waited for me countless times, and I remain inspired by his all-too-short life.

A few weeks earlier and two hundred thousand years after the origin of modern humans on the Dark Continent, we'd crouched around a cooking fire as night fell on Mubwindi Swamp. With Bob Drewes, Jens Vindum, and Jim O'Brien from the California Academy of Sciences, I'd come to the Bwindi-Impenetrable Forest Reserve to help inventory its amphibians and reptiles. Bob is an expert on African frogs, and our findings might bolster the area's protected status, but after a decade of Costa Rican fieldwork I also yearned to satisfy my curiosity about Old World tropical snakes. Jan Kalina, an avian ecologist who ran Bwindi projects with her husband, Tom Butynski, had organized our trip, and several Ugandan wildlife personnel rounded out the group.

Jan and Tom's spare but comfortable field station near the rural community of Ruhizha served as a base, and after two days collecting there our team moved into surrounding uplands. We "four fat boys in the forest" rapidly lost weight thanks to the rugged terrain, but also because Jan, a slim vegetarian, supplied the groceries. Breakfast was always *posho*, a kind of porridge, whereas with few precious exceptions other meals consisted of spaghetti, potatoes, or rice flavored with tomato paste or ground nuts, as peanuts are called locally. Each night I rationed out one thin piece per herpetologist from an Italian salami purchased in the San Francisco airport, and, always hungry for meat, we held the slices on our tongues like fat-soaked communion wafers. At Bob's suggestion we carried bottles of hot sauce for flavoring, and halfway through the trip our cook enhanced the dinner pot with an exceptionally tough and scrawny rooster, purchased along a road that skirted the reserve.

Mubwindi camp was seven thousand feet above sea level on a terrace beside the swamp, and the hike from Ruhizha took several hours, after which late-afternoon exploring left us rain drenched and cold. As dusk fell, pairs of ibis flew low and silently over the water, reminiscent of stealthy jet fighters, then squawked like New Year's Eve horns when they landed. Cicadas screeched incessantly and frogs filled the gathering blackness with boinks, chirps, and trills. I huddled with the others under a tarp shelter and ached all over from paths that had seemed almost vertical as I struggled to keep up with the Ugandans. After I'd had a hot meal my thoughts bounced from noisy turacos and other brightly colored birds to shockingly large elephant dung on the tunnel-like forest trails. I was especially puzzled by the strong body odor, at once strange and familiar, of the game guard I'd followed all day. Neither our ethnic and dietary differences nor his lifetime in the forest explained Vincent's provocative aroma, and at first I couldn't place it.

My hiking companion was the most outgoing of those who guided us. Vincent had a ready sense of humor and was enthusiastic about everything. I was in awe of his wilderness skills. He fashioned rope by weaving strips of sapling bark, made toast out of moldy bread skewered over hot coals, scoured cooking pots with grass, and carried a firearm "to intimidate poachers." Within hours of our first meeting this cheerful, generous man was teasing me about keeping up, then doubled back from hundreds of yards up a hill to show me where gorillas had slept the night before. As we walked on, Vincent told me how game guards roughed up captured poachers, and near the end of our visit he gave me two confiscated spears from a pile stashed in a shed at Ruhizha.

That first night at Mubwindi Swamp I fell asleep imagining Vincent's demeanor as he interrogated prisoners, and the next morning, as a chilly dawn crept over us, I had an unsettling dream that someone was methodically tossing small stones against our tent. Eventually, awakened by the hooting of chimpanzees, I discovered that frogs we'd collected the night before, hopping and softly croaking in a plastic bag next to my head, were the culprits. A few moments later, still lying in my sleeping bag and listening to the low murmuring of the Ugandans while they made tea, I realized that Vincent smelled like thirty years of campfire smoke.

Arboreal vipers exemplify processes that shape tropical biotas, and having studied the Central American species, I now wanted to see their African counterparts. Palm-pitvipers split from terrestrial relatives when ancient oceans overran Mexico's Isthmus of Tehuantepec, and subsequent uplifting, from the Chiapan highlands to Costa Rica's Cordillera de Talamanca, separated coastal and montane prototypes. A Nicaraguan seaway further separated highland lineages, after which earthly rumblings and Pleistocene climate cycles isolated their cloud forest homelands, yielding still more varieties. Finally, about three million years ago, lowland eyelash palm-pitvipers dispersed into South America with the emergence of the Panamanian land-bridge.[7] Today those snakes challenge us to adopt the respectful caution of forest dwellers and behave less like careless intruders. The odd nuance of a mossy patch, idly scrutinized during a lunch break, may reveal catlike pupils surrounded by green coils; among several bright flowers, one might unfurl, suspended by a prehensile tail, and flash an open-mouthed threat display. Could it be otherwise, I wondered, in our australopithecine birthplace?

African bush vipers also evolved during fragmentation of tropical forests, including those of the Eastern Arc Mountains of Kenya and Tanzania, and parallel their Neotropical analogs in several respects. Green bush vipers come in yellow and orange, like eyelash palm-pitvipers, and both gape at predators.[8] But up close and personal, would arboreal Old World serpents remind me of those I knew firsthand in Latin America? Their homeland is the least studied large landmass biologically, even as its habitats dwindle under human impact, and this heartbreaking dilemma was obvious during our trip. We found only six live snakes, whereas Charles Pitman, in his 1974 *A Guide to the Snakes of Uganda,* described the country as "plentifully serpent stocked."[9] Pitman even said Great Lakes bush vipers were abundant in papyrus stands, but for days on end I found no snakes there or anywhere else. Finally one morning we picked up a pregnant rhinoceros adder, her skin resplendent in black, greens, blues, and reds, body as big around as a grapefruit—surely among the most lavishly patterned and stoutest of all serpents. Someone had stoned her to death.

My high point came a few nights later, after an hour sloshing in and out of shallow swamp pools, peering upward and sweeping a light over vegetation, looking for the opalescent reflection of snake bellies. *Frogs, frogs, and more frogs* . . . then even as I muttered "Huh?" my headlamp beam snapped back over its path through the foliage and telltale coils materialized in a sapling ten feet above me. Catching venomous snakes in trees can be tricky, so I took a deep breath and looked off into the darkness for a few seconds.

What would be the smart way to do this?

I made a crude platform of fallen branches, climbed onto it without disturbing my quarry, and looked up to make sure the serpent couldn't fall on my face. Then I delicately grasped the juvenile green bush viper with three-foot-long snake tongs, without disrupting its resting posture, and lowered my prize into a cloth snakebag. After a sigh of relief I called out, "Bob, just got my number one goal in Africa!" And as if that weren't enough, the little bush viper gained luster during a photography session the next afternoon when I noticed its open-mouth display differed in some particulars from that of palm-pitvipers. The newly captured animal extended its head with fangs erect when my camera lens came close, whereas New World species cock their necks back with front teeth folded when they confront predators.[10]

There was another surprise among the few snakes we caught. Nocturnal serpents usually have subdued markings and hide when inactive, but we discovered a thirty-inch, velvety green *Dipsadoboa* with catlike pupils that prowls streams at night and sleeps on foliage by day. It had a frog in its stomach. Perhaps at cool elevations the leaf-colored Bwindi snake must bask even as it sleeps to ensure warm body temperatures for digesting bulky prey, thus requiring a camouflage unsuited to relatives in the warmer lowlands. With this lovely creature in hand, I remembered the nocturnal emerald tree boa of South America, the palm-pitvipers I'd seen in Costa Rica, and pondered anew whether similar ecological demands explain such unusual coloration in distantly related arboreal species.

Whatever the significance of gaping displays and bright colors, throughout human history we have faced serpents similar to those one encounters today. Modern snakes differentiated in the Eocene, more than thirty-five million years ago, whereas the famous fossil Lucy's line diverged from chimpanzees and bonobos in the early Pliocene, about seven million years ago. Stone tools go back two million years, fire use by *Homo erectus* a mere four hundred thousand years, and the first written records are some five thousand years old. As we'll discuss later, early humans, like other primates, surely regarded serpents with a mixture of fear and curiosity, resulting in accurate lore and wild legends. I'll bet bipedal, nest-raiding apes marveled at egg-eaters swallowing objects larger than their own heads, and that those savanna-ranging hominids respected the hissing threat of an Olduvai puff adder. Snakes must have been common in the African Pliocene, just as they still were until recently, and our ancestors were no doubt revering, reviling, and killing them.[11]

On rainy afternoons in Bwindi we preserved specimens and wrote notes, sitting under a tarp on whatever was handy to keep off the muddy ground. One day I pulled a big green and yellow Ruwenzori chameleon out of a collecting bag, placed it on my knee, and admired the three horns on its snout. As the animal crept along with the rocking, hand-over-hand gait characteristic of its clan, Vincent folded his arms and shivered. Disgust hung on his face. Bob had explained that many Africans are extraordinarily fearful of these reptiles, and now, with an air of relief, my new friend exclaimed, "You must take special medicine to handle dangerous creatures!" A few minutes later I gave the chameleon a fatal drug injection and prepared it as a museum specimen, which for lizards and snakes includes forcing formalin into the tail to evert the paired copulatory organs for study. I cradled the dead animal in one hand, inserted a needle behind the penile swellings, and as hydraulic pressure from the syringe caused the two structures to balloon out, Vincent's eyes went wide. Momentarily speechless, he then exclaimed with an envious smile, "Those could be very useful!"

Around midnight we gathered around a small fire, dog tired, the conversation predictable for naturalists on an extended field trip. We talked about snakes and frogs, then traded updates on the successes and failures of our respective digestive tracts. Finally, knowing his culture is polygamous, I said, "Vincent, how do you arrange things with more than one wife? Do you all live in one house?" "Oh no," he chuckled. "They'd fight! Each one must have a *separate* house." Bob caught my eye and nodded, so I pushed on. "How's that work? Do you spend a week at a time in each house, or what?" Vincent's eyes flashed in the flickering light and he replied evenly, "Oh I visit *each* house *each* night." We four white guys, approaching the physical uncertainties of middle age, exchanged skeptical glances, and after a few seconds of letting us hang like that, Vincent burst out laughing.

As the weeks passed, such humorous interludes couldn't conceal an unsettling disparity between first-world conservation activities, especially as conducted in comfortable Berkeley coffee shops or high-powered Washington boardrooms, and the gritty, on-site work of preserving African rainforests. Tom and Jan were respected scientists who hoped to raise a family, yet lived on yearly contracts and traveled hours over rough roads for basic medical care. Vincent's salary covered only part of his family's subsistence needs, and I was humbled that he didn't poach or steal. Throughout rural Uganda, death was casually omnipresent, as if lives were cheap.

One morning, four forestry department officials arrived as we broke camp on the edge of Bwindi, and the scene rapidly turned grim. While one of them in a long coat stood with his assault rifle not quite pointed at us, their leader, a thin, ominous man with high forehead and sparse goatee, launched a vitriolic tirade about affronts to his jurisdiction. I remembered Vincent saying that trespassing soldiers had recently disarmed our game guards, forced them to lie naked on the ground, and then fired bursts from automatic weapons. Thinking the others dead, he'd dashed for the forest and arrived at Ruhizha three days later, incoherent from fatigue and fear. Now those same men regarded the forestry people with smoldering glares. I quietly asked one who was standing next to me and fingering his rifle trigger, "Francis, what's going on?" "They are all corrupt and we hate them," he whispered back. "They hate us too, and sometimes we shoot each other."

We white guys smiled and kept quiet, deferring to Jan. *Hate us* and *shoot each other* clattered around in my brain with *how many rounds are in that Kalashnikov's banana clip?* Later I learned that Bob, like me an army veteran, was also pondering where to dive for cover and how to get one of the guns if a firefight broke out. Instead, the only woman present diffused the situation by acknowledging the goateed man's authority yet just as firmly insisting on the legality of our research permits, and after we all shook hands the forestry people roared away on their motorcycles.

Bwindi means "darkest of dark places," and our journey felt like a descent through human history coupled with apocalyptic visions of the future. It had begun with twenty hours of jet flight, followed by a jolting twelve-hour truck ride across an almost totally converted landscape. As we traveled through neighboring Kenya, Nairobi National Park evoked surprisingly powerful nostalgia for North American prairies. Surrounded by zebras, buffalo, and several species of antelope, I imagined the richness of other equally grand, now extinct savanna faunas: not many centuries back the Great Plains held thirty million bison, and only twelve thousand years ago mammoths, ground sloths, and saber-toothed cats inhabited North America. But a month later, looking south at the Virunga Volcanoes from a forested ridge near the Rwandan border, I had never felt more remote from Western culture, even though we'd seen no undisturbed habitat during the drive from Kampala— only people and livestock everywhere.

On that first night in Mubwindi Swamp campfire fumes burned my eyes, evoking a whimsical phrase from teenage hiking trips. "Vincent," I quipped, "in my country we say, 'Smoke follows beauty.'" After a pause, he grinned and replied in precisely enunciated British English, "Here people say, 'Smoke follows one who *dee-fee-cates* too close to the trail.'" While we looked for snakes in Africa, on foot in the cradle of our primal quest for fire, I sometimes glimpsed a darker irony in those lighthearted comments. With so many paths crowded and foul, I wondered, where will the mingled destinies of snakes and people lead next?

Ten years later, after I'd moved to Cornell University, came the stunning news that Vincent was dead. Following the widely publicized massacre by Rwandan rebels of several foreign tourists near Bwindi, a group of Ugandan soldiers had been garrisoned at the park. One of them abducted Vincent's mother from a bar and raped her. She contracted syphilis from the encounter and sought help at a nearby clinic; when the soldier himself came in for treatment, the episode became public knowledge. Details were sketchy as to what happened next, but the family suffered harassment from the military. Vincent began drinking heavily; he was posted to a ranger camp so far from Ruhizha that he only infrequently went home, and his relationships with coworkers deteriorated. Then one afternoon this man who'd shined with such passion took drunken refuge in a guard hut and ended it all—whether by accident or on purpose was unclear—with the rifle he'd used to protect Bwindi gorillas.

The night after Jan's letter arrived I poured a glass of Merlot and sat down to Simon and Garfunkel's *Concert in Central Park*. Vincent shared a remote forest with elephants and giant vipers; he could have fallen through root tangles into an unseen chasm and never been found. Yet the man owned a confident, guileless innocence, and I'd imagined his life in that romantic vein. Now I remembered that at the time of my visit Uganda was still suffering from horrific excesses of the dictator Idi Amin, and I wondered whether I'd missed something back in 1990, been blinded by Western preconceptions and naive hero worship. Perhaps while admiring Vincent's wilderness lifestyle I had failed to appreciate some latent sorrow in him. Maybe I couldn't recognize it in myself.

I sipped wine and listened to the music, freighted with imagery from my own violent youth, thinking about my recent lecture to a class on creativity. After I'd shown Michael and Patricia Fogden's stunning images from our book about snakes and talked about studying nature, a student in a jaunty orange baseball cap challenged my sympathetic portrayal of reptiles. "I could photograph *serial murderers*," he asserted, "surrounded by children and smiling, but my friend's pet boa *kills* mice, so what's the difference? Aren't your beautiful predators wicked too?" I admitted no special knowledge of philosophy, then said that having been around victims of war and sexual assault, and having had a loved one murdered, I believe viscerally in evil[12]—but surely snakes eating rodents are just doing their job. Now, though, only hours after learning of Vincent's death, the distinction felt pointless. A man who could have been trampled by large mammals or died from myriad other natural dangers had succumbed to something not right within himself, and what weighed most heavily was the loss.

In wilderness there are always matters of danger and the unknown, as well as moments that affect us in surprisingly primal ways. A bushmaster was spotted at dawn near the La Selva lab clearing, and although more than seven feet long and weighing perhaps nine pounds, it was only an average-sized adult of the largest viper. Several of us had abandoned breakfast and stood safely distant—we knew that although this species rarely bites humans, about half its victims, despite treatment, succumb to tissue damage and shock. The bushmaster was rod-straight and crawling slowly, but as more people arrived it lay still, head just off the ground and cocked toward us. About once a second its tongue swept down, out to the sides, and over the snout, measuring our relevance in its morning. Among the assembled researchers were some whose prejudices threaten snakes everywhere, but under the spell of this magnificent creature we all slipped into a sort of naturalists' trance.

After a few minutes the bushmaster began inching toward the sprawling, buttressed trunk of a nearby tree. Two keel-billed toucans landed on an overhead limb and howler monkeys roared from back in the forest, reminding us that the day was under way. One woman pointed out how well the snake's blotched color pattern and keeled scales blended into its forest-floor carpet of leaves and branches, and her companion added that such effective camouflage was all the more impressive given that the animal's midbody was as big around as his arm. A Costa Rican student mentioned its Spanish name, *matabuey,* and asked whether a bushmaster can really "kill an ox." Everyone knew that few biologists have seen this species in the wild, that we were privileged.

I squatted and looked at the serpent's onyx eyes and forked tongue, which serve perceptions plausibly similar to ours. Its pitviper namesakes, like sunken radar screens between the eyes and nostrils, were even more intriguing. One of my early reptile books showed blindfolded rattlers striking hot-water-filled balloons,[13] and subsequent research revealed that sensory receptors in the pits are spatially represented relative to each other by cells in the brain. Spatial patterns of nerve firings thus create infrared images, allowing these snakes to evaluate enemies or prey by combining information from pits and eyes.[14] Perhaps that bushmaster saw me as a dark blob silhouetted against cooler vegetation, and because I neither moved nor came closer, I wasn't a threat. After brief inspection of my size, shape, and surface chemistry, I became irrelevant background. Within minutes the viper was asleep, shaded by a fallen log, and as shafts of sunlight warmed the foliage our handful of bipedal primates set about daily tasks, ever more alert for venomous snakes.

Predator-prey interactions remind us that nature is exuberantly complex as well as brutally indifferent. She is also sexy. One night at La Selva, Carole Farneti and Richard Foster showed *Secrets of the Mangroves,* a film they'd made about Bornean wildlife. Two dozen field biologists had put in a sweaty day's work and filled their bellies with dinner; rain pounded the tin roof as Richard fiddled with the projector and we sipped beers and chatted, a little black bat hawking insects among the rafters only a few feet over our heads. Then we sat enthralled by footage of crab-eating frogs, proboscis monkeys, and other Asian treasures. After half an hour the story line turned nocturnal, and suddenly a cinnamon brown bat, head strangely primatelike, hovered down onto an extravagant white flower. Most of us had learned in zoology classes that the wings of bats are suspended mainly on bones of their hands and thus are unlike those of birds, whose flight structures are supported mostly by their arms. Now, though, without consciously thinking about anatomy, we understood that this gorgeous creature was reaching out for the flower with its *fingers.*

At first the bat awkwardly grasped the curved and densely spaced outer petals, but then its membranous hands delicately enveloped the inflorescence, as one might cradle a lover's face. Next the flying mammal pushed its furry muzzle into the lush blossom, snaked out a reddish tongue, and began lapping nectar with insistent, slaking strokes. Flecks of pollen glistened on its snout. As well-read tropical biologists, we knew that those tiny grains would rub off on the female parts of other flowers, that this bat would inadvertently facilitate plant sex. Our minds reflected on those details as our hearts wandered elsewhere. I heard a murmur or two in the darkness. Someone coughed quietly. The woman next to me undid a ponytail and shook her hair free, then leaned over and whispered, "A third of these people have no idea why some of us are pleasantly agitated. Of the remainder," she said with a big grin, "some of you are fantasizing about being a bat and the rest of us are wishing we were that flower."

Science yields tangible facts as well as more subtle rewards. Looking into a bushmaster's face—staring down those eyes and pits, mesmerized by that flickering tongue—we confront unbridgeable gaps where knowledge gives way to wonder. As spectators to the coevolutionary dance of flowers and pollinators, we are reminded that life's very complexity expands the realms of imagination, that even biochemistry, its much-deserved acclaim notwithstanding, will never fully fathom the cognitive barrier between a viper and us, nor diminish the poetry of a nectar-feeding bat. There are things we will never understand, and as Pablo Neruda said in *Bestiario,* we'll always want to "go farther and deeper."[15] Whether *campesino* or game guard, bird watcher or herpetologist, we need these profound mysteries, above all for drawing us out of ourselves, as reminders that the cosmos doesn't turn on individual lives. To those ends and more, La Selva preserves a spectacular piece of Costa Rican rainforest, and Bwindi has been declared a Ugandan national park. Wild places are still inhabited by bushmasters and jaguars. For now at least, Vincent Bashekura's gorillas are safe.

NINE

Giant Serpents

WE NOW TURN FROM FAVORITE PLACES to their inhabitants: specifically, long scaly ones who challenge our notions of superiority and safety, even our sanity. An official from the Centers for Disease Control once told me, "Hazard equals risk plus outrage," explaining why threats loom out of proportion to their likelihood of affecting us. In that vein, at Cornell University, where I now teach, cars bully pedestrians and a drunken bus driver once ran over a student, yet dogs are banned from campus buildings to protect our "health and welfare." Likewise, despite the fact that the main cause of violent death in the United States has a steering wheel rather than teeth, California, where vehicles kill four thousand people annually, declared mountain lions a problem following their first predation on humans in almost a century. Even naturalists are prone to warped notions of risk: although my most dramatic field injury has been a fingernail kicked out by an armadillo and I've usually stayed calm around huge vipers, bring on a tarantula and watch this arachnophobic reptile lover jump and holler like there's no tomorrow.

However theatrically some humans react to spiders, visceral dread for serpents is more freighted by prospects of an awful demise. After all, they strike from hiding, inflict pain, digest us from inside out, and each year kill upward of a hundred thousand people. Ophidiophobia (from the Greek *ophidion,* snake, and *phobos,* fright or panic) seems rational in countries where cobras and vipers abound—India has about forty-six thousand snakebite deaths annually versus twice that many traffic fatalities—and we might forgive missionary Albert Schweitzer, who famously revered all life but shot African snakes.[1] In the United States, though, with maybe ten snakebite deaths per year, it makes no sense that folks who otherwise value wildlife approve of killing venomous species in parks and slaughtering them in front of children at rattlesnake roundups.[2] Among our supposedly more learned brethren, scientists must strive harder to justify studies of snakes than of birds or mammals.[3] One nature magazine editor even labeled my phrase "appreciating rattlesnakes" oxymoronic.

Fear and loathing in the near absence of danger are all the more puzzling given that serpents enthrall some of us and entire cultures revere them. As children, Henry Fitch and I were neighborhood shamans, wielding harmless snakes before terrified adults and gaining attention from peers. More generally, public reptile exhibits are popular, hundreds of biologists study snakes, and millions of others keep them as pets. Serpents have been carved on bone implements, painted on rock shelters, and sculpted on marble porticos; they represented healing to the ancient Greeks, and for that reason a pair of entwined snakes still symbolizes medicine. Today Hindus, Hopis, and Appalachian Pentecostals incorporate venomous species in sacred rituals, and rural people zealously protect Madagascan ground boas, Texas indigos, and other snakes that eat species we regard as pests.

Beyond emotional satisfaction and symbolism, we have also long used serpents for material benefits. Prehistoric kitchen middens contained eastern diamond-backed rattlesnake bones, and modern subsistence hunters eat African pythons.[4] Entire districts of Asian cities are devoted to processing reptiles for food, snakeskin products sell for hundreds of millions of dollars annually, and the pet trade has reached mind-boggling levels—in one recent year two hundred million dollars' worth of snakes and snake products was imported into the United States alone. And despite claims of promoting safety and research, rattlesnake roundups are fueled by economics, bringing hundreds of thousands of dollars to small Oklahoma and Texas towns in a weekend.

All that individual and cultural complexity raises some obvious questions: Does our ambivalence reflect scriptural linkage with the fall of man? Could skin-shedding signify renewal? And if our attitudes toward serpents reflect something more primal, like predation, why are we fearful of *and* drawn to them? In an earlier book, I noted that snakes kill and are killed by tarsiers, monkeys, and apes, which implies that they have inspired terror as well as curiosity since the very dawn of primate ancestry. From our earliest ponderings, I reasoned, snakes have loomed as horrifying yet edible, paradoxically repulsive and at the same time valuable. Animal behaviorist Lynne Isbell has since advanced that perspective with a more comprehensive theory. Ancestral primates faced constrictors that could eat them, she points out in *The Fruit, the Tree, and the Serpent,* and by the time apes diverged from monkeys, venomous species also had appeared.[5] This chronology corresponds with major events in brain and vision evolution, and as we shall see, new evidence supports her hypothesis that fear of being squeezed predates our dread of toxic bites.

As background, serpents originated at least one hundred million, and maybe as much as a hundred and forty million years ago.[6] Among salient features, elongation—allowing them to enter places unsuitable for thicker bodies—limb reduction, and a forked tongue accompanied the group's debut, and they share with lizards the distinction of having paired penises. Snakes are unusually successful, with about 3,400 living species—more than twice as many as geckos or skinks. Although modern pythons swallow calves and other notoriously large meals, the earliest serpents likely ate animals without backbones, while their crafty ways and chemical worldviews served to explore new habitats and avoid enemies. Those first snakes were irrelevant to us, millions of years too early and too small to be dangerous anyway, although adaptations for eating bulky items were surely in place before primates came on the scene. Through it all, snakes have lain low and been hard to see. And as we'll discuss in the next chapter, venoms are a more recent, profoundly important addition to that lifestyle.

On our side of the adversarial relationship, primates guide grasping hands and feet with big brains and binocular vision, unlike rodents, who steer paws by olfaction. Our particular branch of the mammalian evolutionary tree began almost eighty million years ago, before the decline of large dinosaurs at the end of the Mesozoic. Flowering plants and insects flourished as we diverged from a common ancestor with flying lemurs, our joint pedigree already having split from tree-shrews—themselves squirrel-like creatures with flexible limbs and somewhat forward-facing eyes. From those earliest primates, one basal lineage led to Madagascan lemurs, African galagos and pottos, and Asian lorises; Asian tarsiers are closer to the other main lineage, comprising three groups of anthropoids. New World monkeys and marmosets are climbers, often with prehensile tails, whereas Old World monkeys and baboons are frequently terrestrial. The third group of anthropoids, called hominoids, is more closely related to that last one and includes gibbons, orangutans, gorillas, chimpanzees, bonobos, and humans.

Together these narratives confirm that snakes had been around at least twenty million years when the first, lemurlike primates appeared. By then, as the Mesozoic ended, a prospering lineage of macrostomatan ("big-mouthed") serpents, related to boas and pythons, had combined fine-tuned ambush skills, constricting behavior, and expansive swallowing abilities. The cards were stacked in their favor when it came to eating mammals, but then again ancestral primates were sharp-eyed, agile-handed, crafty—and getting more so by leaps and bounds, looking to make whatever they could of snakes and yet ever more terrified of them. We'd embarked upon an evolutionary arms race that continues to the present, fraught with age-old ambiguities.

Given that nod to deep history, attitudes toward snakes are complexly determined. Decades of experiments and field observations document that whereas some individual primates react to serpents with terror, others are just as resolutely curious or indifferent. Ophidiophobia is indeed easier to induce and harder to extinguish than an aversion to guns, and naive children pick out hidden snakes in photo arrays, suggestive of ancient inborn responses[7]— yet some of us aren't afraid, even as youngsters, and in any case experience also shapes phobias. On balance, our australopithecine ancestors must have been attuned to natural history, their ophidiophobia and ophidiophilia reflecting a mélange of innate tendencies, personal travails, and group-held traditions. By the time those bipedal apes began roaming Pliocene African savannas, roughly four million years ago, interest in and dread of snakes were already acute, preceding the use of fire, art, and language.

Lynne Isbell's theory is supported by research showing that we recognize and attend to danger with a cluster of sequentially evolved brain innovations. As she emphasized, snakes originated prior to raptors and carnivores, so constrictors may have facilitated the origin of a uniquely mammalian fear module—a small, deep-seated region between the temples that is controlled by the amygdala and sends more links to cognitive centers than vice versa. It's no coincidence when terror trumps reason! Moreover, primate brain areas specialized for vigilance and alarm connect with neural networks sensitive to the elongate shapes and partially hidden, repetitive geometric patterns typical of snakes. However much our feelings derive from culture and experience, Lynne concludes, we preconsciously evaluate serpentine features thanks to some truly ancient circuitry. With all that anatomical baggage in mind, this and the next chapter summarize the biology of my favorite animals, in hopes of enhancing appreciation for their many admirable attributes.

Snakes, as pop novelist Tom Robbins wrote in *Still Life with Woodpecker,* are "all belt and no pants":[8] they traverse highly textured, even hostile landscapes

Curious primates and deadly serpents: humans (Graham Alexander and author, *right*) watching a ninety-seven-pound, fourteen-foot southern African python at Dinokeng Preserve, South Africa, September 25, 2011. (Photo: K. R. Zamudio)

with humbling ease, and limbless locomotion lies at the heart of their intrigue. Sir Richard Owen, a nineteenth-century anatomist, claimed that they "out swim the fish and out climb the monkey,"[9] and in biblical Proverbs "the way of a serpent on a rock" and that of "an eagle in the air" are "too wonderful" for comprehension. Snake movements are as perplexing as flight, but whereas we envy birds, no curious Leonardo da Vinci ever designed a slithering machine. Nonetheless, although crawling is a dubious prospect and so costly as to appear improbable, it's a common theme over the evolutionary long haul. A more detailed look reveals how snakes have achieved such spectacular success, along with tangible examples of their influence on primates.

The origin of tetrapod limbs left us with about thirty thousand species of living amphibians, reptiles (including birds), and mammals, yet appendages have been *lost* surprisingly often in their nearly four-hundred-million-year history. Of course, there aren't limbless birds or frogs, arms and legs being the essence of flying and jumping, but the amphibian relatives of frogs include superficially wormlike caecilians as well as sirens and several other near-limbless salamanders. Most impressively, limbs have undergone independent reduction and loss more than a hundred times just among lizards, repeatedly within skinks and certain other groups—and one of those evolutionary experiments led to snakes. Extant species with reduced limbs invariably have elongate bodies, implying that over and over tetrapods first got skinny, then gave up appendages as other travel modes evolved.[10] Over geological time, the giant python that might eat you today came from a small, insectivorous lizard with tiny legs.

Locomotion entails support, propulsion, and steering, accomplished without limbs in serpents and best appreciated while watching an undisturbed snake moving in nature. At a bare bones level, limbless reptiles are stretched-out versions of more conventional tetrapods. Drop girdles, arms, and legs, multiply vertebrae and ribs about twenty-fold, and you've got a snaky human. Whereas our skeleton totals 206 bones, an average snake boasts about that many vertebrae, and then there's the ribs and skull elements. (Pythons, boas, and a few others also retain a tiny pelvis and vestigial hind legs, called spurs.) Serpentine muscles resemble in anatomy and function those tissues that move our ribs and backbone, but some span up to thirty or more vertebrae, in overlapping, weblike fashion. As many as twenty or more repeated sets of muscles on each side tug on vertebrae, ribs, and skin, thereby effecting different, often continuously changing movements along the body.

What are the reasons for this radical shift by which limbs were repeatedly lost? Skinny organisms move in ways so distinctive they're given special names: "laterally undulating" in symmetrical waves, "sidewinding" by lifting themselves forward, crawling "rectilinearly" like caterpillars, and pulling and pushing through tunnels and up trees in "concertina" fashion. Reduced diameter also permits entering crevices, requires less force for burrowing (shove a tent stake into soil, then try a piece of fence post of the same length), and allows passage over frail substrates (backpackers fall forward in quicksand, spreading out their weight while crawling to safety, whereas serpents skim across foliage that couldn't support a running squirrel of the same mass).[11] And as observed in our black-tailed rattler field studies, snakes, unlike stoutly built lizards, can leave a body loop in the sun and warm an abdomen swollen with food or embryos even while hiding from predators under a rock or log.[12]

Costs are the other side of the skinny-animal coin, two of which involve energy. First, just as our arms and legs cool rapidly due to high surface area, a snake-shaped mammal couldn't sustain high body temperatures. A weasel, the closest approximation to an endothermic serpent, barely hangs on by eating nonstop, whereas ectotherms warm up by basking, have lower metabolic rates, and simply hide out when it's cold. Second, slender amphibians and reptiles have narrower heads than stout ones do, yet have to gain equivalent nutrition. They've solved this little-mouth dilemma by eating lots of little things, eating chunks of big things, or reorganizing their heads and eating big things intact. Accordingly, blindsnake stomachs contain hundreds of tiny insect larvae, caecilians shear off earthworm pieces, and some snakes consume prey exceeding their own weight, a stupendous feat I'll explain shortly.

Avoiding being eaten poses a third problem for skinny creatures because, all else equal, greater length makes detection more likely, and disabling blows can fall anywhere along a slender body. As for solutions, like most other animals, reptiles generally avoid discovery, flee if found, and confront enemies only as a last-ditch gambit. Judging by their closest living relatives, early snakes were secretive burrowers with drab colors for camouflage when they crawled on the surface, and if attacked their writhing struggles exposed brightly barred bellies that mimicked unpalatable millipedes and centipedes. If harder pressed by a predator they expelled noxious secretions from tail glands uniquely characteristic of serpents, whereas venomous bites and more elaborate defensive tactics came later in their evolution.[13]

One other attribute deserves emphasis. Although a forked tongue evolved repeatedly among lizards, most obviously in monitors and tegus, only snakes have a throat skeleton so simplified that it no longer assists swallowing and functions solely for lingual movement; only the lightly built and exquisitely controlled snake tongue waves about repeatedly during protrusion, its tines together or widely splayed, tips curved upward or down; and only a snake's tongue delivers airborne molecules to corresponding right and left parts of a supersensitive vomeronasal organ in the roof of its mouth, thereby providing directional cues.[14] Thanks to rapid volleys and sustained flicking of this remarkable structure, that favorite blacktail of mine mentioned earlier routinely tracked her bitten prey dozens of yards, all the while ignoring the abundant spoor of other mammals. And thanks to the same extraordinary capacity for processing complex chemical information, male 26 followed superfemale 21's every turn over far longer distances—but only in years when her trail advertised sexual receptivity.

The first serpents perhaps used their fancy tongues and palatal chemo-sensors to search for insects, as do many modern lizards and blindsnakes, but their descendants soon shifted to heavier items. And just as modern pipesnakes eat eels nearly equaling themselves in weight, ancestral snakes could have radically changed foraging economics without increasing gape. Whereas lizards consume many arthropods daily, each hunt-and-capture bout exposing them to danger, proto-pipesnakes could have cut that to a few risky forays per year by pulling their heads over *long slender* prey. And during that early lifestyle shift, regardless of details, snakes likely subdued their adversaries with a limbless version of the bear hug—a small boa tightened around one's arm elicits more discomfort than a blood pressure cuff, and can dispatch a rat in about a minute. Constriction in turn facilitated the evolution of macrostomatans, able to swallow items shaped more like hamburgers than hotdogs. One six-foot-long, fifty-million-year-old fossil "big-mouth" I studied contained a two-foot crocodile weighing 45 percent of its own mass.[15]

How, then, does an anaconda eat a caiman, or a python an antelope? After all, even their average meals are tantamount to me swallowing a ninety-pound object without benefit of hands or utensils.

Eating big prey obviously involves anatomy, including highly elastic skin and ligaments, but one oft-recited characterization is simply wrong. Rather than unhinging or disjointing their jaws, early serpents invented a method for ingesting bulky prey by *dropping* one hinge and *adding* others. In most vertebrates, lower jaw halves fuse during development at a joint—felt as a groove between bumps on our chins—but that suture never forms in snakes. Their ancestors also freed up some skull bones and stuck in extra joints, so the lower jaws hang from struts rather than articulating directly with the skull. And while vertebrate upper jaws are usually firmly attached and support a continuous curved row of maxillary teeth, early serpents united them with other bones from the palate into movable right and left upper-jaw arches. Each arch is shaped like a forward-pointing tuning fork and armed with two rows of sharp, curved teeth that prevent prey escape. Muscles anchored to the underside of the skull insert from different angles on each fork handle to pull the arches forward, backward, up, down, and to the side.[16]

Snake jaws amount to right and left vertically operating tongs, held wide open when prey is snared with teeth. Each pair consists of an upper jaw arch and a lower jaw, such that during swallowing the arches alternately pull the head over prey while the lower jaws stabilize it. You can visualize this by clasping hands in front of you, with elbows at the sides, imagining your chest and forearms as a lizard head. Swing your forearms down and up, simulating jaws opening and closing, and note how the triangular hole bounded by the "skull" and "jaws" limits morsel size. Now unclasp and move your hands apart, as if the lower jaws were free of each other at the chin; then extend them halfway at the elbows and angle your upper arms out like struts. That larger open space accommodates a serpent's bulkier meal, and hand-over-hand movements mime side-to-side lower jaw actions. Or put more simply, imagine hooking index finger- and thumb-tips into a stocking gathered at the ankle, then pulling this makeshift snake head—right hand, left hand, right hand, left hand—over the food item represented by your bulging calf.[17]

Thus equipped, when primates appeared on the scene, snakes were ready and waiting.

Only a few living species of serpents are longer than twenty feet and thereby qualify as giants. These include the green anaconda, reticulated python, Asian rock python, and northern African python. Scrub pythons and three other poorly known Australian and New Guinea species may also attain that length, but are more slender than the others.[18] Two species occasionally exceed thirty feet, but Bronx Zoo's long-standing offer of fifty thousand dollars for a live one that size has never been claimed, and until recently knowledge of their biology has been correspondingly skimpy.

Early explorers routinely exaggerated snake size, as with a green anaconda purportedly one hundred feet long and weighing five tons,[19] and even modern coverage is often more objectifying than revealing. Robert Twigger's *Big Snake,* for example, tells of a British writer, about to marry and yearning for one last adventure, who learns of the Bronx Zoo reward and sets off in search of a thirty-foot reticulated python.[20] Twigger engagingly recounts the discomforts of working in rainforests, and his descriptions of Asian people feel authentic, but the title's namesake never inspires empathy, even as an adversary. What little herpetology we get is outdated—and perhaps that's the problem—since the oft-quoted C. J. P. Ionides was a superb hunter who above all knew how to find African snakes. But Hollywood portrayals are where hyperbole goes off the charts, as in *Anaconda,* where the film crew and maniacal boat captain square off against a colossal snake depicted with hilarious biological fabrications.

Despite the ongoing prominence of giant serpents in myths, explorers' tales, and horror movies, until recently there were no Fitchian studies of the big ones. Henry's own tropical work centered on parts of Latin America where anacondas were rare at best, and like many other snake biologists he never made it to the Old World tropics. Even there, giant serpents are hard to find and difficult to handle. Nonetheless, a renaissance in snake natural history is well under way, in which traditional approaches are extended by research on dead animals and wild ones are implanted with radio transmitters. Although emphasis has been on gartersnakes, adders, and other common small species, knowledge of the largest snakes is growing rapidly.

Reticulated pythons occur in southeastern Asia, Indonesia, and the Philippine Islands. Although a recent photograph from Laos shows one said to have measured thirty-six feet, the accepted maximum has long been thirty-three feet, and one that size might exceed two hundred pounds. Even such big retics are fairly slender as pythons go, with distinct necks, angular heads, and a netlike tapestry of rainforest hues that conceals them from enemies and prey. Their brown dorsal blotches are irregularly connected and set off from a lighter gray or yellowish background. Smaller black- and yellow-edged spots descend onto the flanks as bars or Y-shaped markings, and each color conglomerate encloses a white patch. Dark stripes run back from the orange, yellow, or gray eyes and disrupt overall head shape. Retics are reputedly mean-tempered, and an adult's teeth can sever the nerves, tendons, and blood vessels of a good-sized primate.

In the 1990s Richard Shine, my Australian alter ego—about the same age, likewise bald and bearded, unabashedly enamored of reptiles—conducted the first life history studies of giant snakes.[21] His team measured more than 1,800 retics in Sumatran tanneries, then examined reproductive organs, gut contents, fat bodies, and parasites. Local people brought the pythons from fields, plantations, forests, and villages, having caught them on hooks baited with rats or chickens and in snares along trackways. Slaughterhouse workers killed the snakes with a wire poked in the brain, a blow to the head, or suffocation, gruesome nuances no doubt lost on stylish buyers of snakeskin boots and purses. Carcasses were suspended by the neck and filled with water to facilitate processing, after which skins were removed and pegged for drying, meat was removed, and gall bladders were dried for use in folk medicine.

Rick discovered that male retics breed at six feet and rarely reach fifteen, whereas females mature at a third longer and may exceed thirty feet—consistent with a trend among snake species for longer females to produce more offspring. Retic clutches average twenty-four eggs, each four by two inches and weighing half a pound, and large females lay more than seventy eggs. Hatchlings start at thirty inches and eat rodents. Small adults feed on rats and chickens, but giants take monitor lizards, pigs, deer, porcupines, pangolins, civets, and monkeys; one had eaten a 132-pound hog! Pythons coexisting with denser human populations in northern Sumatra exhibited greater size at maturity, larger eggs, more frequent feeding, fewer parasites, and smaller fat stores than their southern cousins. Big pythons, however, were scarce in the north, reflecting poor survivorship around people as well as abundance of rodents, more appropriate prey for smaller snakes.

German doctoral student Mark Auliya further illuminated the lives of Indonesian and Malaysian pythons by visiting tanneries, using funnel traps, and sifting the observations of other naturalists and local people. Retics are terrestrial and arboreal, favoring the vicinity of streams and other wetlands, especially swamp forests; primarily nocturnal, they repeatedly use caves and other shelter sites, yet may travel half a mile in a month. Females coil around their eggs during sixty to ninety days of incubation, in a hollow tree or under a fallen log. Besides the rodents, deer, pigs, and monkeys that prevailed in earlier diet studies, Auliya recorded bats, tarsiers, and a binturong, the largest civet. One eighteen-foot retic contained a long-tailed macaque and her clinging infant, and a twenty-two-footer beside a stream was digesting a barking deer.[22]

Gabriella Fredriksson scored another fascinating glimpse of big snake natural history when a 130-pound Borneo retic ate the fifty-one-pound radio-collared sun bear she was tracking. Having confirmed that her study animal was active one afternoon, the Dutch conservation biologist located its telemetry signal the next morning in a thicket a third of a mile away, emanating from the bulging twenty-three-foot serpent. Upon being poked with a stick the python retreated to a stream, remained there four days, then moved four hundred feet and took refuge in an enormous hollow log. Twenty-six days later the python returned to the stream; it was captured a week later and kept captive until it defecated bear bones, after which the transmitter was surgically removed.[23]

Knowledge of the heaviest living snake species also accumulated slowly, through both random encounters and, more recently, targeted field research. Green anacondas inhabit savannas and rainforest in the Orinoco and

Amazon drainages. Dark spots punctuate the murky tan to dull green skin, such that these huge boas are shockingly obscure in their watery lairs. The head is smaller and more rounded than in pythons, its outline disrupted by black-bordered orange stripes, and the eyes and nostrils are elevated for lurking beneath the water surface. Green anacondas reach twenty feet in the llanos of northern South America and likely much longer in Amazonia—one report of a thirty-seven-footer feels reliable, especially given the recent discovery in Colombia of a sixty-million-year-old, forty-two-foot fossil relative![24] And the big gals are undeniably massive. Male green anacondas usually reach less than ten feet in length, but a twenty-foot female in prime condition might exceed three hundred pounds, a thirty-foot snake twice as much. The fossil behemoth, aptly named *Titanoboa,* would have weighed in at well over a ton.

Green anacondas like lagoons, swamps, and streams, often basking nearby or hiding in waterside holes and vegetation. Females deliver twenty to eighty two-foot-long neonates and devour the birthing debris and stillborn young, thereby recouping energy and eliminating odors that attract predators. Observations and dissections have revealed an unusually broad diet of fishes, frogs, caiman, turtles, lizards, anteaters, capybaras and other rodents, peccaries, deer, foxes, dogs, cattle, diverse birds, and the occasional human. Anecdotes further hint at a rich array of interactions with other animals. A green anaconda in Surinam snatched up a wattled jacana as the sharp-eyed bird waded over vegetation,[25] and an Internet film clip shows an enormous Brazilian individual disgorging a capybara, its jaws astonishingly far apart on the carcass.[26] Black-chested moustached tamarins in Peru often crossed a lake inlet on fallen trees, until an anaconda struck from the depths and ate one of the shaggy little critters.[27] For days the survivors avoided that route, and perhaps their descendants still chirp nervously around water.

One of my academic younger brothers, Jesús Rivas, obtained a Ph.D. with Gordon Burghardt studying green anacondas on the seasonally flooded Venezuelan llanos. For seven years he conducted visual surveys, probed barefoot for submerged snakes, and followed dozens with radiotelemetry, obtaining data on hundreds of individuals with a passionate, swashbuckling flair that plays well in wildlife documentaries. Spectacled caiman, capybara, and white-tailed deer were often eaten, and although herpetologists had long ridiculed claims that constrictors break their victims' bones, anacondas, it turns out, do just that. As for causes of mortality, mammals, caimans, and bigger anacondas take juveniles, whereas large females succumb to disease, overheating, and injuries from prey. One perished with a turtle wedged in her throat, and another was fatally impaled by the catfish she swallowed.[28]

Jesús's team also documented the giant water boa's mating aggregations, each of which consists of several yards of writhing coils and looks like a cornucopic Medusa heaped up in the shallows. A dozen or so smaller suitors entwine along an enormous, loosely coiled or stretched-out female for up to forty-six days, during which they stroke her body with pelvic spurs and struggle with their hind parts to father her offspring. All indications are that this is as complex as it looks, and through it all females seemingly control whom they let in, maximizing reproductive success by choosing one or more of the best males. Obviously we still have much to learn about snake social systems!

Just as birth certificates and bank records don't adequately describe us, statistics on reproduction and diet fall short of fully revealing other animals' natural history. What *is* it like to be a giant snake? Reticulated pythons and green anacondas resemble their close relatives in many ways besides ancient constricting behavior, so perhaps vignettes of other pythons and boas can help us envision lifestyles of the big and famous. African pythons, for example, eat vervet monkeys, who in turn adopt vigilant postures and give specific alarm cries at the sight of one—yet they react to python tracks with curiosity and jump back only when they find the large snakes themselves in the grass. When biologists interrupted a ten-foot Madagascan ground boa constricting a Coquerel's sifaka and freed the lemur, they noted that she had enlarged nipples and refused to leave the vicinity, then saw a bulge in the snake's foreparts, presumably her baby. And as with anacondas, things don't always go well for other big serpents; a sixteen-foot Asian rock python died when a ninety-pound hog deer with fifteen-inch antlers lodged in its gullet.[29]

Turning to a widespread relative of anacondas, boa constrictors are superbly camouflaged in habitats ranging from Mexican thorn scrub to Amazonian rainforest. The intricately repetitious colors of one I found at La Selva were nearly indistinguishable from the creases, splotches, and curves of its woody basking site. These eclectic hunters ambush lizards, birds, rats, deer, even ocelots and anteaters. One Panamanian individual moved among mammal burrows every few days, waiting for their occupants' return, and another in Surinam caught three cock-of-the-rock males at a courtship site, the birds' numbered leg bands later showing up in the snake's droppings. A Costa Rican boa killed a white-faced capuchin while its comrades screamed and threw sticks; the snake hissed and struck but didn't abandon the dead monkey. And a Peruvian five-footer seized a subadult black-chested moustached tamarin from low on a tree in which it fed, whereupon a male and female bit the captor's head and neck while four others vocalized. The victim escaped and licked its wounds when the snake retreated into a hole, and its group avoided that tree for a couple of days, then fed higher when they returned.

Of course, all is not "kill or be killed." A six-foot common tree boa lay draped over a liana in Peru, her symmetrical coils a cryptic patchwork of browns and yellows in the sun-flecked rainforest understory, her chunky head oddly doglike. For hours, a smaller male flicked his tongue up and down the female's neck while vibrating pelvic spurs against her skin, then coiled nearby, after which they lowered hind parts and he wrapped his tail around her gaping vent. Later, several saddle-backed tamarins chattered and mobbed the copulating snakes until they slowly withdrew into a dense vine tangle, entwined tails still hanging downward.[30] We have no idea how often or why primates harass snakes, but our interactions are obviously multifaceted— even more so with the big ones, as we shall see. Meanwhile, the prize for a thirty-footer remains unclaimed, and a book that captures the majesty of giant serpents has yet to be written.

Despite groundbreaking research and a trove of anecdotes, three aspects of big snake biology that inspire widespread curiosity remain poorly studied. First, although green anacondas are heaviest, and they or retics are longest, we don't know their true maximum sizes. Second, because the behavior of caged snakes likely has about as much resemblance to that of wild anacondas and pythons as the antics of a farm pony do to those of feral mustangs, we have little basis for imagining the daily lives of giant serpents. A third topic is tied to the others, because size influences what can be subdued and swallowed: Under what circumstances do snakes eat people, and thus affect us psychologically and culturally? A couple of former Cornell University students helped illuminate these puzzles and presaged my own first encounter in nature with a really large snake.

Our acquisition of one of the largest serpents displayed anywhere began when Reed McJunkin, class of '32, contacted Cornell officials about a skeleton in his closet. Reed's father had worked in the Philippines from 1908 to 1916 with the U.S. Army's Bureau of Insular Affairs, developing schools in northern Luzon. Norman McJunkin was an avid hunter, and on one trip with fellow officers, the first night out, they built a campfire under a tree.

Their Filipino porters were terrified of snakes, so in response to commotions overhead, Norman fired his shotgun two or three times and something crashed into the nearby darkness. No one dared investigate until daybreak, when they discovered a twenty-six-foot reticulated python; they laid the carcass among ten-foot-high anthills and continued hunting. Scavengers devoured the python's flesh within days, after which Norman collected its bones and brought them back to the United States with other souvenirs. Over the next four decades, "Ralph" resided in his Pittsburgh and New York homes, where McJunkin kids and grandkids relished laying the skeleton out at holiday gatherings.

Reed McJunkin wanted to donate Ralph to his alma mater. I called his grandson Scott Palmer, and after some pleasantries—as a boy I'd lived on Luzon; Scott had been there in the navy—he agreed to ship their herpetological heirloom. Next I enlisted David Cundall and Frances Irish, Lehigh University snake anatomists, to assemble the skeleton. It had been stored as disarticulated ribs and skull bones, with the vertebral column strung on a cord, and for dozens of hours we sorted more than one thousand pieces of an osteological puzzle. David and Fran laid the backbone out in their living room, then painstakingly checked 357 vertebrae for proper positioning. We separated ribs as right or left based on the bumps with which they fit against vertebrae, estimating their placement along the spinal column by size and shape. A third of the skull bones and more than a hundred ribs were missing, but our labors led to a spectacular exhibit nonetheless. The finished specimen is about twenty feet long—reasonably close to the original measurement, given loss of soft tissue among the vertebrae and decades of shrinkage.

I like to think Norman would have chuckled to learn that Ralph was a female, as are all really large boas and pythons. Thanks to the McJunkin family's generosity, their big snake skeleton now occupies an elegant cherrywood case at our Museum of Vertebrates, where it attracts many admirers. Few giant serpents in museums have accurate collecting data; now researchers have access to one that grew up under particular natural conditions, rather than in captivity. We discovered, for example, gnarly evidence of healed fractures on some fifty ribs, as well as along Ralph's backbone. Whether enemies or prey caused these impressive injuries, they further document that even big snakes endure rough times. As such findings are combined with field studies, a reasonably complete picture of their natural history will emerge.

The second Cornell connection addresses in spectacular fashion whether giant serpents can be a serious threat to humans, a topic of long-standing debate. Venomous snakebites are common among Third World agrarians and forest-dwellers, but records of constrictors eating people are rare—one southern African python fatally ambushed a teenager on a path, and reticulated pythons kill several rural Asians every year. Green anacondas sometimes attack humans who are bathing or fishing, and once a big female that struck from hiding narrowly missed Jesús Rivas's collaborator María Muñoz.[31] One bizarre saga involves a photograph I first saw in 1998 of a nineteen-foot Sulawesi retic opened to expose the clothed body of a man.[32] Years later an identical image appeared on the Internet, labeled as depicting a Venezuelan oil worker who'd gone off to relieve himself and ended up inside a green anaconda. The fiction subsequently became absurd in a newspaper photo of the same python with the same man, captioned as an anaconda that had eaten a Bolivian woman! Given a lack of evidence to the contrary, herpetologists have generally concluded that snakes scarcely ever attack people as food.

In 1999, I received a spectacular photograph of Philippine Negritos with a retic, twenty-three feet long and twenty-six inches in circumference. It came through Timothy Wright, a professional geographer with long-standing interest in giant reptiles. Tim's daughter Amber took my first herpetology course at Cornell, and I met him at her graduation. Shortly thereafter he sent me the photo, which he'd carried around for almost three decades. Together, we then set about locating the photographers.

Anthropologists Thomas and Janet Headland began studying the Agta Negritos in 1962, became fluent in their language, and assembled unparalleled evidence of snake predation on humans.[33] Negritos were widespread until recently, though now they are marginalized as peasants and threatened with extinction. When Tom and Janet first arrived in Luzon's Sierra Madre, the Agta were still preliterate, lived in small kin-related groups, slept in tiny temporary shelters, foraged in old-growth rainforest, and ate wild meat daily.

Tom surveyed 120 Agta about their relationships with pythons; his respondents were sixteen to seventy-five years old, and their lives encompassed many encounters. The big snakes had attacked fifteen of the men but only one woman, presumably because the men spent more time in the forest. Fourteen respondents were struck once and two of them twice, totaling eighteen attacks; fifteen were actually bitten, presumably as a prelude to constriction, and eleven had scarred lower limbs to prove it. Pythons had struck along trails and usually were dispatched with a bolo knife or homemade shotgun, but between 1934 and 1973, according to Tom's informants, six Agta were killed, including a grandfather, an uncle, an acquaintance, and a woman who died

Philippine Negritos with freshly killed reticulated python, twenty-three feet long and twenty-six inches in circumference, at the headwaters of the Koso River in the Sierra Madre, Aurora Province, Luzon, June 9, 1970. (Photo: T. & J. Headland)

two months after being bitten on the leg. In the most recent incident, the snake entered a thatched hut at sunset, constricted two children, and was swallowing one of them headfirst when the father intervened and dispatched the python with his bolo. Years later, the Headlands interviewed the mother and her daughter who had survived the incident, as well as an American Catholic priest who buried the other two children and observed their bite wounds.

Slow-moving, heavy-bodied snakes depend on camouflage and active defense to avoid enemies, rather than crawling away. Once discovered, giant serpents are easily killed with simple weapons, and not surprisingly (because snakes don't have toxic flesh), people turn the tables and eat them too. Tom observed that the Agta routinely hunted and consumed Philippine deer, Philippine warty pigs, long-tailed macaques, and reticulated pythons, but not other snakes or domestic pigs. All adults had probably killed at least one python, and occasionally they encountered large ones. The two hunters in the photograph skinned and butchered that snake in less than an hour; assuming that 33 percent of her mass was usable, they thus obtained about fifty-five pounds of meat for their group. Other Agta discovered a thirty-two-pound wild pig in the stomach of a twenty-one-foot python they killed.

Tom's Agta were not in a general sense "primitive," but his data provide a rare estimate of predation risk for hunter-gatherers and refute a widespread misconception. Herpetologists had often claimed that snakes don't eat humans, but after all, indigenous people in Africa, Asia, and South America are usually smaller than Western scientists. At 60 percent of a retic's weight, a hundred-pound Agta man would not be heavy, especially for a serpent whose prey includes 130-pound pigs. Moreover, the eighteen unsuccessful attacks and six fatalities amount to a traumatic incident in the group every year or two, and deaths would have been more common a few centuries ago, before the advent of metal weaponry. If python predation approached the incidence of unsuccessful attacks on Agta, it would have exceeded the 8 percent of all male deaths that forest-dwelling Paraguayan Aché incur from jaguars.[34]

Anthropologists have long argued that predators influenced human evolution, and some recently discovered australopithecine bones bear tooth marks from crocodiles and mammalian carnivores.[35] As for snakes, although their few fossilized stomach contents haven't included primates, Tom's data on hunter-gatherers similar in size to Lucy and her kin are unequivocal. Pythons frequently attacked Negritos, and vice versa; moreover, because big snakes observably ate deer, pigs, and monkeys, the humans would have regarded them not just as predators and prey, but also as competitors. Until recently, the Agta's lives were indeed complicated by multifaceted, dangerous ecological relationships with giant serpents. Moreover, as we have seen, other primates prey upon and are killed by snakes—over its lifetime an individual retic might well eat tree-shrews, tarsiers, macaques, gibbons, and people—all consistent with the notion that a shared ancient heritage has inspired our dueling attitudes toward limbless reptiles.

About the same time as those Cornell connections were established I met an exceptionally large snake while visiting the Bronx Zoo with my herpetology class. Samantha was a reticulated python, wild caught as an adult and almost twenty-three feet long. She had an enormous head, perhaps for having in youth eaten larger prey than her captive fare of forty-pound piglets, and given inclination and opportunity she surely could have swallowed me. I looked through the glass cage front and imagined her back in Borneo, moving rectilinearly like an immense caterpillar over dank forest substrate. What would it be like in her world of chemical and tactile sensations, I wondered, with one's head so far from one's tail? And out in nature, how would her presence affect other creatures?

I lacked field experience with any snake like Ralph or Samantha until one evening in 2007, when I entered a Mato Grosso swamp with Kelly and her Brazilian research collaborators. We were collecting frogs, but my mind drifted to the following week, when we would tour a flooded savanna near the Bolivian border. The Pantanal is famously rich in wildlife, and since we'd have wetland access by boat and horseback, I hoped for a yellow anaconda; we were on the edge of the much heftier green anaconda's geographic range, but even its smaller cousin might surpass the largest wild snake I'd ever seen. That ten-foot Costa Rican boa constrictor, though impressively surly, was hardly a giant compared to some pythons and anacondas.

Now I waded in knee-deep water, camera slung over one shoulder, and consigned a treefrog to the plastic bag clutched in my left hand. Fernando Zara, an invertebrate zoologist whose ophidiophobia rivals my discomfort around his beloved arthropods, walked several yards ahead, and Natália Pansonato, a Brazilian undergraduate who had never seen a live snake, was off to Zara's left. The others were across open water behind a palm grove, barely within earshot, and like us they searched for chorusing amphibians and swatted insects in the heavy night air. Our headlamps sliced back and forth through the swamp, its vegetation glistening under silver clouds and a poignantly full moon.

My light passed along branches, up and down yard-long grass leaves, as I scanned for frogs to catch and spiders to avoid. Every minute or so I glanced about for roots that might snag a boot or a bottom hole that could send me splashing, all the while humming a Mark Knopfler song about "digging up a diamond, rare and fine."[36] Suddenly there were staggered, dark coffee saucers winding past my leg, like black leaves twisting in an olive current, and after a flash of dissonance—*lily pads should be on the surface, and why are they moving?*—my brain shouted, *Green anaconda!* As her tail came into view I grabbed with one hand, whereupon the snake jerked loose, reappearing a couple of yards to my right. I took a few steps and lightly laid hands on the massive torso, worried that restraint would provoke constriction, and waited for it to taper. Then, after feeling the dimple of her vent pass over a fingertip, I seized the tail and pulled up toward my waist.

The next few seconds were chaotic, heavy with snaky unknowns. Zara sloshed over to help, but my only plan was to drag the anaconda onto land, get a better look—and make another plan. The snake's torso glinted like a submarine parting the moonlit waters, then arced sideways, and I wondered if big toothy jaws were swinging our way. Instead the snake strained forward, sliding against the plastic bag in my left hand, and I struggled backward with the tail clutched in my right hand as, inexorably, she pulled free. Zara and I circled in the shallows scanning for her spots, but my spider-loving friend lagged back, and I flinched as something bumped my leg, so maybe visions of formidable girth had compromised our enthusiasm. The two of us were babbling—"She was huge!" I shouted, to which Zara replied, "Puta merda, Harry, ela era enorme!"—and Kelly called for quiet because they were recording frog calls. When I looked toward where we'd walked in off the road, Natália sat wild-eyed and silent atop a fencepost, as if treed by some phantasmagorical swamp monster.

Zara and I cross-checked estimates of the distance between nearby bushes, then agreed that the anaconda was at least fifteen to twenty feet long, because we observed more than four yards of body, with no narrowing of foreparts. We never saw her head. Three inches separated my hands on each side as the trunk slid between them, and when I grasped the tail my thumb and forefinger were still more than an inch apart. The skin was slick and tough, neither cold nor slimy, and she moved with muscular confidence, as if I were only a momentary distraction. A few weeks later, back in wintry New York, hauling her out of the water seemed about as plausible as pulling my pickup truck up our icy driveway.

No yellow anacondas greeted me in the Pantanal, but thanks to that lunar evening I more keenly empathize with those Peruvian tamarins, veering wide of where days earlier an explosive lunge snatched one of their band into oblivion. Now I better appreciate how a green anaconda may appear as benign leaves and stones on a pond bottom, why a deer or a monkey, or for that matter a person, at water's edge won't detect her stealthy presence. There will be no sound as she approaches, no scented warning from upwind, and sinuous ripples in the swamp grass might not be a storm's breezy prelude. Now, without consciously thinking about such things, I understand why complacency doesn't bode well in the home of giant serpents.

TEN

Venomous Serpents

THE LITTLE SOUTH TEXAS CEMETERY FEELS melancholic despite limpid blue skies, its somber mood reinforced by a windmill's creaking dirge somewhere off in the thorn-scrub. I've come to check out a lichen-smudged tombstone, weathered by more than a century of sun, wind, and occasional rain, that reads "John D. Sweeten, born in Denton Co. Nov 14, 1862, was bitten by a snake July 16 in Atascosa Co., died in San Antonio July 24, 1880." Ecologist Larry Gilbert, who grew up near here and alerted me to the grave marker, reckons it was a two-day wagon ride to the nearest doctor—where remedies might have included mind-numbing doses of whiskey, incision and suction, kerosene, and a fresh-killed chicken split over the wound. Unrelenting, excruciating pain was a given, gangrene likely, so the teenager's death from the bite of a western diamond-backed rattler would have been traumatic for all concerned. No wonder keeping ranch yards well swept was all about scaring away six-foot buzz-tails, as was having vigilant pets and tolerating rattler-eating indigo snakes around houses and schools.

Constrictors launched our fear of serpents in the late Mesozoic, almost eighty million years ago, but venomous species such as the one that killed Sweeten have surely posed greater risks during the second half of primate history.[1] For one thing, the seventy-four extant species of boas and pythons generally inhabit tropical regions, and usually strike at prey rather than in self-defense. By contrast, nearly ten times that many *dangerously venomous* species—defined as those that might kill humans—range from Scandinavia to Patagonia, from Death Valley to the high Himalayas, and although most are too small to swallow anything larger than a squirrel monkey, their biochemical arsenals can incapacitate a gorilla. Furthermore, when venomous snakes feel put upon, they flagrantly challenge our supremacy. Boas and

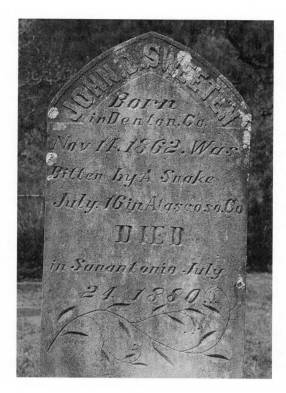

Tombstone of nineteenth-
century snakebite victim,
Atascosa County, Texas.
(Photo: H. W. Greene)

pythons rely on camouflage, biting only when closely threatened, whereas
coralsnakes flash bright colors and cobras elevate spectacled hoods, adders
hiss loudly and pitvipers rattle their tails, all as if to say, "Are you feeling
lucky, primate? Get bitten and make my day!"

Published vignettes support this scenario of venoms fine-tuning ophidiophobia, by showing that primates do kill and are killed by venomous
snakes. Natural history anecdotes describe, for example, a western tarsier
chomping on a long-glanded coralsnake, white-eared marmosets harassing
jararacas, and white-faced capuchins dispatching a terciopelo with sticks.
Conversely, a jararaca struck a cotton-eared marmoset; black-necked spitting cobras, black mambas, and gaboon adders have eaten bushbabies.
Moreover, defensive bites by venomous snakes likely have always been more
common than predation, as exemplified by a puff adder and a green mamba
killing patas monkeys and Cape cobras biting chacma baboons—in all cases
primates way too large to have been attacked as prey. Lab studies comple-
ment those field observations by revealing brain mechanisms for vigilance

and fear that respond to serpentlike diamond and cross-band patterns, and that, like our anthropoid proclivity for pointing at objects, are plausibly associated with detecting dangerous reptiles. Among mammals, while opossums, hedgehogs, and woodrats evolved immunity to venomous bites, monkeys and apes countered snake toxins with manual dexterity, visual acuity, and cognitive skills.

To what extent venomous serpents affect human primates, negatively or positively, depends on the local scene. Today folks who coexist with western diamondbacks wear boots and have easy access to hospitals, whereas barefoot tropical hunter-gatherers still rely on folk remedies to contend with several species of common dangerous snakes—so although bites are rare in the United States, some Amazonian Amerindians suffer injury rates exceeding those of python attacks on the Philippine Agta. An astonishing 95 percent of Ecuadorian Waorani males had been bitten at least once by vipers or coralsnakes, almost half of them more often; equally surprising, bites caused only 4 percent of deaths in that population, compared to 7 percent from infanticide, 10 percent from illness, and 45 percent from "speared by other Waorani."[2] Snakes understandably terrify forest dwellers and rural people, the more so for their having watched loved ones endure cruel deaths. The tables do get turned, of course: prehistoric middens contain bones of eastern diamond-backed and Middle American rattlesnakes, and worldwide we consume millions of serpents annually.

Given such tumultuous history, it's no surprise that herpetologists are often asked, what's the deadliest snake? Judged by *drop for drop toxicity,* Australia's inland taipan beats all others hands down, with enough toxins in one bite to dispatch two hundred thousand mice. That species, however, rarely bites anyone. Which species *kills the most people* depends on toxicity, amount injected, and snake abundance multiplied by snappiness, victim's size and health, and medical care—such that each year a few species of vipers and cobras fatally bite tens of thousands of Africans and Asians. In the United States, about eight thousand are bitten annually, 70 percent by rattlesnakes, 20 percent by copperheads, 9 percent by cottonmouths, and 1 percent by coralsnakes; fewer than a dozen die, usually bit by rattlers that were purposely handled. Were he bitten today, John D. Sweeten would likely recover without permanent injury, but in some countries snakebite remains as much a threat as it likely was for early anthropoids.

Statistics, of course, don't matter when the snake materializes next to your hand or foot, poised to defend itself. And if it does strike? Viper bites typically destroy tissue, sometimes causing loss of digits or even limbs, and in severe cases massive blood leakage leads to fatal shock. Neurotoxic cobras, kraits, and their relatives are another matter, because although local symptoms usually aren't as dramatic, without treatment respiratory failure is likely. Most formerly recommended snakebite first-aid measures are useless or worse—as a Boy Scout I dreaded carving Xs over fang marks—and because antivenin works well, prompt transport to a clinic is typically the preferred option. One of the world's premier snakebite experts told me, however, that if bitten by a large rattlesnake and needing more than a couple of hours to reach help, "I'll deny this if quoted, but I'd apply an arterial tourniquet." When I recalled learning to use that risky measure only for severe bleeding, he countered, "Quite so, but you see, I'd be trading limb loss for death from shock." And I once met a Russian zoo curator who, having no gaboon adder antivenin, immediately removed his bitten thumb with a hatchet kept nearby for just that purpose.

Epidemiologists label bites *illegitimate* for folks who purposely interact with snakes, and these occur more frequently than *legitimate* accidents. Scars and missing fingers were traits of the "herpetologist handshake" when I first attended scientific meetings in the 1970s, and professional deaths were well known. By the mid–twentieth century a terciopelo had killed Douglas March at his Honduran serpentarium, a spectacled cobra fatally bit zoo curator Grace Olive Wiley during a magazine interview, and physician-naturalist Fred Shannon lapsed into irreversible shock from a Mohave rattler bite. In 1958, the boomslang that Field Museum's Karl Schmidt was asked to identify ended his life, and after a 1975 bite from his pet savanna twigsnake, Senckenberg Museum's Robert Mertens uttered the last words "What a fitting death for a herpetologist." One can only sadly wonder whether a fourteen-inch Burmese krait, seemingly too small to be deadly, would have killed California Academy of Sciences' Joe Slowinski in 2001 if expat Hans Schnurrenberger's rapid demise from the bite of a foot-long MacClelland's coralsnake, thirty-five years earlier in Nepal, had been publicized.[3]

Schmidt succumbed on my thirteenth birthday, and I'd met Mertens in Germany, but until 1981, despite their deaths, I experienced only admiration during many run-ins with a dozen kinds of pitvipers, several coralsnakes, and a couple of European adders. Up to that point I'd never worked where snakes *felt* like serious threats, but during the ensuing decade in Costa Rica two species afforded some viscerally thought-provoking moments: bushmasters and terciopelos could scarcely be more different, but out on a trail at night, the biggest western diamondbacks I'd ever caught couldn't hold a candle to either of them as an electrifying presence. Since then I've met countless other venomous snakes—from golden lanceheads and urutús in Brazil to puff adders and a Cape cobra in South Africa—and even though bumping an unseen bushmaster with my foot was momentarily unnerving, only March's killer had me chronically on edge. Long misnamed *fer-de-lance,* French for "spearhead" and more appropriate for a Caribbean island pitviper, the Mesoamerican species is now called *terciopelo,* Spanish for "velvet," by gringos and campesinos alike—an allusion to the lush appearance of its huge black-and-gold females.

My uneasiness around terciopelos was first inspired by a severe bite sustained by a Berkeley ecologist colleague, Rob Colwell—he'd been struck in the shoulder while walking along a streambank and waited several anxious hours for rescue[4]—and that apprehension was soon reinforced by the inch-long fangs and copious venom flow I encountered when handling them. Even more ominous was pioneering Costa Rican snake expert Clodomiro Picado's sobering description: "Moments after being bitten, the man feels a live fire germinating in the wound, as if red-hot tongs contorted his flesh; that which was mortified enlarges to monstrosity, and lividness invades him. The victim witnesses his body becoming a corpse piece by piece; a chill of death invades all his being, and soon bloody threads fall from his gums; and his eyes, without intending to, will also cry blood, until beaten by suffering and anguish, he loses the sense of reality. If we ask the unlucky man something, he may still see us through blurred eyes, but we get no response; and perhaps a final sweat of red pearls or a mouthful of blackish blood warns of impending death."[5]

Although bushmasters and terciopelos are equally well hidden in their lairs, the former lay a few large eggs and take rodents throughout life, while the latter bear dozens of young that feed first on frogs and lizards, then switch to birds and mammals as they grow larger. Bushmasters attain at least nine feet, prefer primary rainforest, and rarely strike people, whereas terciopelos average five feet, thrive in disturbed areas, and, consistent with their legendary feistiness, cause many snakebites.[6] The first big terciopelo I tried to catch arrayed herself in a striking coil when touched with a snake hook, so that, stuck knee-deep in swamp mud, I fervently wished for a pistol during the few tense seconds until she moved away. A six-footer slithered between my legs as I tried to restrain her, and two others reversed direction while crawling under palm leaves, as if to ambush their pursuers. Despite biologists being bitten only about once every half million person-hours of fieldwork at La Selva, Picado's grim scenario often resonated during my time there, and terciopelos remain the most impressively dangerous snakes with which I have ever worked.

Serpents evolved venoms not to kill pursuers like us but to subdue and digest prey. Their defensive threat displays protect harmless species as well, as in the case of the coralsnake mimics I studied in graduate school. Nonetheless, something sinister lurks beneath all that rationality. During decades of work with venomous species I've suffered only the mild copperhead bite as a teenager, but in 2003 Kelly heard explosive rattling underfoot, saw the blacktail and two puncture marks on her shin, and endured a torturous recovery. The very day I learned of young Sweeten's tombstone, my herpetologist-rancher friend David Hillis reached under some hay, expecting tiger salamanders, and jerked back from a western diamondback—then within thirty minutes leapt over another one by his porch! Snakes, it turns out, remind even those of us who love and study them of how, by evolving new ways to make a living, they've shaped our own, distinctively primate biology. I'll argue shortly that we've influenced theirs too.

At least half of all serpents possess venoms, defined as injected toxins—themselves organism-produced poisons, which are chemicals deleterious to life—and browsing any of herpetology's continental-scale treatises will provide a sense of the dangerous-to-humans fraction of overall snake diversity. In Campbell and Lamar's epic *Venomous Reptiles of the Western Hemisphere,* for example, we can savor images of the yellow-bellied seasnake, 72 species of coralsnakes, and 117 species of pitvipers; lingering over intricate color patterns and individual species' natural histories, we can learn that coralsnakes and lanceheads occupied South America long before wild dogs and cats crossed the Isthmus of Panama onto that continent.[7] Comparably rich snake assemblages populate Old World landmasses. Because venoms are mostly about food (stay tuned for exceptions), we'll discuss three general aspects of getting a meal, before exploring global snake diversity.

First, to maximize their bottom line, predators should pass up prey only if, during the time required to eat it, something better could be taken. "Better" has to account for effort spent searching, catching, and swallowing, as well as predation risks and opportunity costs (such as not mating right then), all relative to food value and, ultimately, reproductive success. Foraging optimally thus amounts to making cost-benefit trade-offs in gaining an adequate diet. Indigo snakes and other wide-foraging searchers may daily travel hundreds of yards, for example, whereas bushmasters and other sit-and-wait ambushers can occupy one site for weeks, during which taking small prey might not be worth the risk of attracting enemies. Generalists often are jacks-of-all-trades but masters of none, like coachwhips that eat grasshoppers, lizards, baby birds, mice, and even small rattlesnakes, but not large centipedes. The plains black-headed snake, a dietary specialist, takes only those formidable myriapods.[8]

Second, because snakes swallow food intact, shape matters. Prey can be elongate and cylindrical, like worms or eels; noncircular in cross section, like laterally compressed fish and animals with awkward protuberances, such as birds; or bulky, like rodents or eggs. For snakes of equal gape, elongate cylindrical prey yield more food than other shapes (hotdogs weigh more than meatballs of equal caliber); a heavier prey item, however, can fight back and, if also bulky, might rot before stomach juices seep into its interior—a serious problem at high latitudes and elevations because an ectotherm's digestion slows down at lower temperatures. Eels are easily swallowed and especially nutritious for their diameter, whereas fish and fowl require more gape for the same calories. High-payoff mammals require large gape, safe dispatch, and rapid digestion. The insects eaten by blindsnakes and most lizards don't require much gape or special handling, but they provide little value per item. For other snakes, feeding infrequently on heavy prey means less exposure to predators—assuming the prey can be efficiently subdued, swallowed, and digested.[9]

The third general point concerns a hunter's own prospects of being dispatched by a predator. As potential prey, a snake might be so cryptic as to impose intolerably high search costs, and/or appear so well defended as not to warrant the energy costs and risks of capture. To avoid enemies, venomous serpents are usually well camouflaged and back up threats with toxic bites, whereas mimics capitalize on the presence of those dangerous models and lie about their own defensive abilities. This all amounts to a high-stakes game, because hungrier predators will pay higher costs, and availability of alternative prey varies. The natural history literature confirms that even extremely dangerous animals aren't invincible—honey badgers and snake eagles have killed black mambas and puff adders, an ocelot mortally wounded a five-and-a-half-foot Totonacan rattlesnake—yet most predators, when they can, take the easy way out, as in red-tailed hawks that will attack western rattlesnakes but more often take nonvenomous gopher snakes.[10]

Given that background, serpent evolution amounts to a three-act play, with plot twists imposed by climate and habitat change, foraging economics, and the origin of novel traits. If Act One was the Mesozoic rise of stout, big-mouthed constrictors in the Age of Reptiles, described in the last chapter, Act Two began with early-Cenozoic expansion of cooler, drier grasslands during the Age of Mammals, when horses and rodents were on the make and slender, supple ancestors of advanced snakes posed an alternative to earlier macrostomatan lineages. By analogy with modern boas on the one hand and racers on the other, those first ambushers survived on a few large meals per year, whereas the stripped-down innovators were out poking, peering, and tongue-flicking into smaller creatures' hideouts. Items that fought back much wouldn't be worth the risk of detection by bigger predators, and depending on prey abundance, diet specialization might have been difficult. On the plus side, advanced snakes could rapidly slither away from their foes, which was not an option for more heavy-bodied species.[11]

Act Two featured another subplot, in which the slimmer, faster models freed up their jaws, perhaps to more quickly ingest struggling prey that otherwise might attract snake predators. Using my analogy of "toothy tuning forks" for the upper jaws of macrostomatans, recall that right and left arches alternate in forward-backward cycles when muscles from the palate tug on them. Most importantly, as individual bones, analogous to the fork prongs, evolved mobility, their inner tooth rows proved adequate for jaw-walking over prey, thereby freeing the outer tooth rows for specialized new roles. Consistent with that shift, dental adaptations are modest at best among boas, pythons, and other remnants of early serpent evolution—emerald tree boas, suspended by their tails from branches, secure prey with long front teeth— whereas advanced snakes include folding-toothed skink-eaters, needle-toothed slug-eaters, club-toothed crab-crushers, and species with diverse styles of venom injection.[12]

Act Three began at least thirty-five million years ago with diversification of vipers; then, five to ten million years later, it continued with radiation of the cobra family, stiletto snakes and their relatives, and—except in Australia, where cobra relatives predominate—roughly two thousand species of advanced snakes that mostly are not dangerous to us.[13] Those harmless serpents include, among others, Northern Hemisphere ratsnakes, racers, and kin; North American watersnakes and gartersnakes, along with Old World counterparts; and a diverse group, including hog-nosed snakes and a few more in the United States, that's primarily found in the New World tropics. Many species are venomous to prey, though unless defined in terms of humans, it's impossible to draw a crisp line between those that do and don't have toxins. In any case, dangerous species and their mimics make up more than half of all serpents. Inventing venoms in the early Cenozoic was clearly at the heart of serpentine success—among other things, it facilitated again feeding on large prey—and one paleontologist facetiously renamed that era the Age of Snakes.

As for the anatomy of venom delivery, imagine forward-directed hypodermic syringes, in which the barrels—situated behind the eyes on a snake's head—serve as production and storage sites for toxins that are forced by plungers through needles into prey or adversaries. At a primitive level, it's all about modifying the glands, muscles, and teeth typical of many vertebrates, and each of the components of this syringe analogy varies among lineages, from vaguely obvious in some species to extremely sophisticated in vipers and cobras. We'll begin with glands because, although many species with venom lack enlarged fangs, there are no instances of the reverse, suggesting that toxins originated before the specialized teeth and muscles that are used to inject them.[14]

To simplify a lot of fascinating research: all serpents have glands along the jaws that secrete mucus and enzymes, compounds that, like those in our saliva, mainly lubricate food and catalyze biochemical reactions. Toxic head glands use evolutionarily transformed versions of those secretions mainly as prey tranquilizers—some work slowly, others within seconds—and/or tenderizers, although some act as spreading factors, pain inducers, and chemical signposts for finding bitten prey. Venom glands consist of toxin-secreting cells packaged in a sheath of cellophanelike connective tissue. In simplest and most widespread form, they lack a lumen (storage chamber) and compressor muscles. Boomslangs and a few other dangerous rear-fanged species, however, have the gland cells arranged in folded rows, such that space between them forms storage tubules, as well as small muscle attachments that may move venom to the teeth. Front-fanged snakes, in contrast, have a well-defined lumen, a duct to the fang, and modified jaw muscles that pump toxins into other animals. Most impressively, although stored venoms remain potent, they don't tranquilize or tenderize the snakes that make them.

Isolated dry fangs are ivory white, but in live snakes they're translucent, reminiscent of fine crystal. Rear fangs are situated behind two or more smaller teeth on the maxillary bones; single, paired, or in triplets, with or without grooves, they are embedded with side-to-side jaw cycles, penetrating tissue as each is pulled down and back, like a fishing gaff. Venom flows under low pressure, from jaw compression and capillary action along the rear fangs, and prey is held until dead or struggling weakly, then swallowed. Front fangs arose at least three times when the maxillaries shortened during embryonic development and moved enlarged rear teeth forward, evolutionary transitions during which grooves closed to form canals for high-pressure injection. Front fangs are either short and fixed or long and folded, depending on whether the bones to which they're attached move slightly or a lot; like hypodermic needles, they have beveled, usually elliptical tip openings. Cobras and other snakes with fixed front fangs typically hold prey until ingestion starts, whereas vipers, with folding front fangs, seize frogs, lizards, and birds but strike and quickly release mammals, then relocate them before swallowing.[15]

We know little of what venoms do in nature, despite countless studies of their biochemistry, but tranquilizing and tenderizing prey surely are key roles. Rear-fanged plains black-headed snakes kill centipedes within minutes, and desert nightsnakes subdue reptiles up to half their own weight; Puerto Rican racers more rapidly digest envenomed lizards, and separate glands of Brazilian snail-eaters tranquilize or tenderize mollusks. Aside from toxic bites, hog-nosed snakes use huge rear fangs to puncture toads that swell with air to thwart ingestion. Only front-fanged serpents, however, immobilize truly heavy prey, illustrated by prey/predator weight ratios of 1.4 for an eastern coralsnake that ate a glass lizard and 1.6 for a common lancehead containing a whiptail. And as predicted if venomous digestion is important where warm seasons are short, outside the tropics, only vipers commonly eat heavy, bulky prey; the northernmost snakes are Scandinavian adders, the southernmost are Patagonian lanceheads, and Himalayan pitvipers and Mexican dusky rattlesnakes live in some of the world's highest mountains.

The biology of venomous serpents seemed straightforward when I studied them in high school—coralsnakes, rattlers, and their kin kill prey and sometimes us; mildly venomous rear-fanged snakes evolved from nonvenomous species, which in turn gave rise to dangerous forms with fixed front fangs and thence, most recently, to folding front fangs. Now, with better understanding of the entwined roles of development and evolution, we know the toxic serpents began flourishing tens of millions of years ago, as advanced snakes loosened those toothy tuning forks and modified their head glands, converted enlarged hind teeth to grooved rear fangs, and repeatedly invented front fangs. Researchers are increasingly focused on the diversification of venomous serpents, mimics, and their adversaries—the players in Acts Two and Three—and I believe we'll soon be asking if toxins have more to do with defense than heretofore realized.

Big-mouthed constrictors had our shifty-eyed, tree shrew–like ancestors to themselves for the first several million years of serpent-primate conflict, but today descendants of those earliest macrostomatans are prominent only where advanced snakes are rare or lacking. Cuba has a nine-foot relative of the South American rainbow boa and fifteen species of dwarf boas, whereas no other comparably sized landmass has more than five total from those two groups; Australia is dominated by one highly specialized bunch of venomous snakes and also boasts twenty species of pythons, whereas all of Africa, with an otherwise much more diverse serpent fauna, has only four of the latter. Most continents are populated instead by hundreds of species of advanced snakes, many blatantly venomous. Because Greek or Latin names denote their lineages, a little taxonomy will help us appreciate them.

The 308 species of Viperidae (Latin, *vivus* and *parus,* to give live birth—as do many vipers, rather than lay eggs) branched off soon after advanced snakes originated and best exemplify the hypodermic syringe analogy. Their short maxillary bones are hinged with the front of the skull, so as the upper jaws push forward, both fangs—an inch or more long in gaboon adders—rotate to almost 180 degrees for high-pressure venom injection. Striking, biting, and releasing take less than a second, during which the snake adjusts fang placement to avoid hitting the victim's bones and dispenses venom according to victim size. Today vipers, having arisen in Asia, are diverse in tropical and semiarid regions; sometimes they are among the largest local predators and remarkably common. Most are terrestrial, but prehensile-tailed arboreal species have evolved independently in Asia, Africa, and the New World, as have side-winding desert forms. African night adders with upturned snouts seem to be the only burrowing viperids, North America's cottonmouth the sole aquatic one.[16]

Vipers are generally stout-bodied ambush hunters, sometimes luring frogs and birds by wriggling a caterpillarlike tail. The young often eat frogs and lizards, whereas adults mostly take mammals. I observed one little western rattlesnake eat a western fence lizard that slightly outweighed it, and an adult had swallowed a California ground squirrel equal to 50 percent of the rattler's premeal weight; the rodent's whiskers were still intact but its abdomen was open, evidently digested from inside out by deeply injected venom.

Peringuey's adder and several other small species eat lizards throughout life, while the large neonates of eastern diamond-backed rattlers and urutús take adult rodents. Emphasis on tranquilizing versus tenderizing venoms parallels predation on lizards or mammals, exemplified by their relative importance for juvenile and adult western rattlers and by geographic diet variation in Malayan pitvipers and massasaugas. Some saw-scaled adders are unusual for having evolved a predilection for scorpions and toxins specialized for killing them.

Vipers show fangs, hiss, rub scales, or vibrate tails prior to striking at enemies—so, despite signature feeding adaptations, defensive novelties also have played key roles in their evolution. Moreover, infrared-imaging pits, whose neural input is integrated with vision, characterize 213 species (69 percent of all vipers), and surprisingly, although capable of prey detection, a pitviper's namesake organs provide no known hunting advantage compared to pitless Old World relatives. These snakes do, however, emphasize risk avoidance and a "good offense is the best defense" strategy: superbly camouflaged, they never rely on striped patterns with which some other snakes facilitate locomotor escape, and their pits evolved right along with protection of offspring and defensive tail vibration. Later, the Mexican ancestor of thirty-seven species of rattlesnakes (17 percent of pitvipers) morphed its tail spine into a string of castanets, perhaps to warn off the probing snouts and paws of carnivores in talus habitats—and today their defensive displays can be heard from Canadian bogs to Argentine thorn-scrub.[17]

The 351 species in Elapidae (Greek, *elop,* a kind of serpent) have fixed front fangs (half an inch long in king cobras) and usually neurotoxic venoms. Typically slender, elapids are more diverse in size, shape, and ecology than vipers; they include secretive serpents no larger than ordinary earthworms, fifteen-foot king cobras that construct nests out of vegetation and guard their clutches, and a seasnake so specialized on fish eggs that its venom apparatus is vestigial. Nine-foot black mambas are the epitome of terrifying, but small, banded, burrowing snake-eaters are prevalent across most elapid lineages, so the group's overall common ancestor likely resembled the one hundred species of Asian and New World coralsnakes in appearance and natural history, including responses to predators. Although these secretive creatures have poor vision, step within their tactile realm and things quickly become chaotic. A restrained coralsnake writhes its curled, elevated tail back and forth, then the real head emerges from a confusion of

Curious primates and deadly serpents: young savanna baboons watching an emaciated black mamba during a severe drought in Amboseli National Park, Kenya, October 13, 2009. The snake's small hood is displayed defensively, and it crawls fast but not frantically. (Photo: C. L. Fitzpatrick)

red, black, and yellow coils, snapping at everything in reach. Within seconds, all hell breaks loose, a would-be captor is bitten, and the gaudy serpent escapes into leaf litter.[18]

Some thirty species of cobras are generally larger than coralsnakes, with prominent eyes and quick responses; most are terrestrial, although Africa hosts a few arboreal, aquatic, and burrowing species. They generally reach at least four feet in length, sometimes twice that; most lay eggs and prey on fish, frogs, lizards, snakes, birds, and mammals, though the rinkhals, which averages about three feet and eats mainly toads, delivers live young. Trademark hoods vary from long and narrow in African forest cobras to short and wide in Asian species, and whereas some African cobras have a dark bar on the underside, Asian spectacled and monocled cobras are named for their hood markings. Frightened cobras generally attempt to flee, but if cornered, perhaps due to such iconic visual displays, they seem more ready to engage us than vipers. And of course there's that bone chilling word, *neurotoxic:* a cobra charges, toxins flow, and we suffocate—possibilities one dwells upon, even if subconsciously, when faced with that telltale flattened neck.

Some cobras upped the confrontational ante by evolving spitting, once in Asia and perhaps twice in Africa. Spitters have round discharge orifices above the fang tip and eject venom streams for six feet or more, prior to which they track an antagonist's head movements and accurately aim for its eyes.[19] Being squirted causes intense pain and, if not treated, blindness. A long-popular theory was that trampling ungulates prompted the evolution of spitting, but in fact herds don't exist in Southeast Asia and spitters arose prior to their advent in Africa. More likely players would have been well armed and had forward-facing eyes, such as swiping, sharp-clawed carnivores and bipedal, tool-using primates—so, keeping in mind reports of capuchins killing a terciopelo with branches and Old World monkeys standing to peer at snakes, and having seen video of a chimp shaking vines and a stick at an approaching python,[20] I'll bet the only long-distance weaponry among all serpents evolved at least partly in response to our anthropoid brethren.

In Australia, it's as if advanced snakes were told to start over, but strictly as cobra relatives and with little competition from other lineages. More than a hundred species of elapids in the Land Down Under encompass brown snakes that look disarmingly like New World coachwhips and also eat frogs and lizards; black mamba–like coastal taipans that likewise feed on mammals and readily attack their enemies; coralsnake-like bandy-bandies that specialize on blindsnakes as prey; and other examples of convergent evolution. Seven species of stout-bodied, wide-headed death adders are remarkably similar to vipers in body shape and behavior, even equipped with slightly mobile fangs and distinctively colored tails with which they lure prey. Australian elapids also spawned two invasions of the sea, including a few species of eel-eating sea-kraits, which lay eggs back on land, and several dozen species of seasnakes with diverse diets, all viviparous and so specialized for swimming that they can scarcely move out of water.[21]

Atractaspididae (Greek, *atraktos,* a thin shaft or arrow, and *aspis,* a viper) comprises about seventy-five species of African and Middle Eastern serpents. All are burrowers, almost all of them rear-fanged and feeding on centipedes or secretive reptiles (quill-snouted snakes eat only worm-lizards), though one relative of African centipede-eaters lacks rear fangs and eats earthworms. Two species of harlequin snakes, however, so closely resemble coralsnakes in having fixed front fangs, colorful writhing defensive displays, and a diet of limbless reptiles that they were long classified as elapids. Likewise, some dozen species of folding-front-fanged stiletto snakes were once referred to the

Viperidae, but their maxillary bones rotate on ball-and-socket joints rather than hinges, such that they stab sideways and backwards with one fang or the other at nestling mammals—and are impossible to hold without being bitten! Front fangs thus evolved once or twice within this small, geographically restricted group, convergent on those of the other two more speciose, dangerously venomous snake lineages.[22]

Advanced snakes that aren't viperids, elapids, or atractaspidids formerly were lumped in Colubridae (Latin, *coluber,* a serpent), whereas now that name is restricted to about seven hundred mostly harmless species, such as the European smooth snake, African egg-eaters, and North American ratsnakes, as well as the rear-fanged brown treesnake and deadly boomslang. Among those no longer called colubrids, most of them venomous and some nearly front-fanged, are hog-nosed snakes and roughly seven hundred neotropical species assigned to Dipsadidae (Greek, *dipsas,* a serpent whose bite causes thirst), along with just over two hundred North American gartersnakes and their kin, a handful of European and African species, and Asian keelbacks, all in Natricidae (Latin, *nato,* to swim; *natricis,* a watersnake). Tiger keelback bites occasionally kill people, and when threatened this species displays a flattened, red nape, where poisonous skin gland secretions, sequestered from toads in its diet, are ejected into an attacker's mouth. Tiger keelbacks also pass toad toxins across their placentas, so young snakes are protected at birth.[23]

Viperidae, Elapidae, Atractaspididae, Colubridae, Dipsadidae, Natricidae, and more—now let's contemplate thirty-five million years of venomous snake evolution in the context of the last thirty-five years of snake research. My dissertation showed how constriction allowed ancestral serpents to subdue heavy prey, favoring evolution of big mouths and digestion of large items, whereas defensive displays are more recent, convergent adaptations to particular lifestyles. Hundreds of people now study snakes, and recent findings imply some additional, overarching patterns involving mobile-jawed, advanced species. The best-known new generality is that venoms, defined by natural history instead of human mortality, are ancient, widespread, and diverse. Moreover, the rich variety of toxins and injection equipment facilitates feeding on many kinds of prey under a wide range of circumstances, rather than representing linear stages toward a single pinnacle of snake evolution.

Another, less widely recognized pattern is that front fangs radically changed prey-predator dynamics in both directions. Boa and python teeth snare prey for constriction, and rear-fanged species must hang on to inject toxins. Thus, if target animals are dangerous, a snake's range of prey and feasibility of defensive biting are limited. Vipers, elapids, and stiletto snakes, however, swiftly dispatch prey *and* punish attackers, making possible not only heavy, infrequent meals, but also the effective use of threats as predator deterrents. Among many examples from the field, several times the buzz of an unseen terciopelo or golden lancehead, vibrating its tail on leaves, has stopped me cold and likely prevented an accident. I'd be surprised if venoms aren't modified for defense in spitting cobras, and I can't help wondering if some other snakes need all that toxicity just to immobilize prey. I'm also impressed that so many Waorani survived bites, as do most domestic dogs,[24] allowing the victims to modify their behavior in future run-ins with snakes. And the evolutionary significance of front fangs doesn't end with feeding and defending their bearers.

To better appreciate the third pattern—that many snakes are mimics— note that snakes similar in both appearance and ecological niche usually are on different continents. New Guinea green tree pythons and South American emerald tree boas, for example, independently evolved to look like foliage on tree limbs and ambush prey with dangling head postures. Conversely, when distantly related snakes look alike and live in one place, they routinely have different niches—the coralsnake mimics I studied eat amphibians, not other snakes—and one of the lookalikes is front-fanged. Many mimics resemble dangerous models in one or two aspects, like head shape and general color pattern, but some are so precisely similar that even herpetologists are cautious. On a field trip to Vietnam, I almost lost an escaping spotted catsnake for fear of grabbing the more commonly encountered and similarly patterned Chinese habu, and in Costa Rica we always treated toad-eating false terciopelos as the real deal until discerning their round, unviperlike pupils. In fact, entire specialized lineages, totaling many dozens of species, closely resemble vipers, cobras, or coralsnakes. Absent venomous models, there'd be no toothless scale-rubbing egg-eaters, no head-spreading toad-eaters, no amplified-hissing gopher and pine snakes, and so forth.[25]

Finally, serpents have mattered to us since the very origin of primates, and conversely, our lineage likely has influenced them far more than traditionally acknowledged. We've shared almost eighty million years of killings, bona fide threats, mimetic bluffs, retaliatory weaponry, and cultural traditions—impelled by constriction and front-fanged venom injection in their corner, by binocular vision, weapon-grasping hands, and cognitive superiority in ours. That evolutionary dialectic likely intensified on the Plio-Pleistocene savannas of Africa—the famous Laetoli and Olduvai hominid sites have yielded fossil spitting cobras and puff adders, one of which had eaten a hare[26]—and still challenges our efforts to coexist with snakes.

Two problems hinder appreciation of serpents, one being our long history of mutual strife and ophidiophobia, the other ignorance of their lives as individuals. The fact that Henry Fitch never saw a copperhead feed in the course of all his Kansas fieldwork, and that I've watched only a handful of snakes eat during decades outdoors, underscores the second point. How can we illuminate the daily comings and goings of such low-key, secretive creatures, the better to understand and care about them? As a teenager, I viewed with dismay *National Geographic* images of radio-collared grizzlies, because their telemetry transmitters were so blatantly inappropriate for animals with no shoulders and a propensity for tight squeezes. By the 1970s, radios were miniaturized for snakes, and although early studies with force-fed units were crude—one researcher sewed thread through the foreparts to prevent regurgitation—keep in mind that the Wright brothers didn't build a space shuttle. Over the ensuing decades telemetry revolutionized snake biology, and the surprises keep coming, some of them powerfully relevant to conservation.

In the summer of 1980 I manually restrained a western rattlesnake in Berkeley's Strawberry Canyon and pushed a paraffin-encased radio the size of an unshelled peanut down its throat, then used a Citizens' Band receiver to follow the beeping rattler and watch it ambush, trail, and eat a field mouse. Observing snake behavior in the field was so instantly addictive that I soon journeyed to Costa Rica, intent on radio-tracking pitvipers at La Selva. Obvious downsides were the risks that manual restraint posed to both the snakes and me, and the passage of transmitters out one end or the other, often within days of being ingested. Nonetheless, we soon documented a juvenile bushmaster hunting three, six, and twenty-four days at different sites, moving some fifty yards total. On the thirty-third night she ate what we inferred was a rodent equaling 40 percent of her weight, based on the shape and size of the food lump; she then remained in place for nine days before moving on. A seven-foot adult lingered for three weeks at one site, resting by day and hunting by night, before leaving us its shed skin, and a scat full of spiny rat hair with the reusable radio.

Technology ramped up with publication in 1982 of a surgical implantation technique by grad student Howard Reinert and his advisor, David Cundall, the anatomist who later helped assemble the giant Cornell python skeleton.[27] Shortly thereafter, Tucson physician Dave Hardy—fascinated by bushmasters since his army brat childhood in Panama—joined me, and thanks to his expertise we abandoned manual restraint and ingested transmitters for the Reinert-Cundall method. First, I'd gently coax a pitviper into a plastic tube, in which it was exposed to the same vaporizing anesthetic used on people. Then Dave snipped a small incision, behind the stomach and forward of the gonads, through which he inserted a lipstick-size radio. Finally, the radio's ten-inch antenna was threaded forward under the skin, and a few sutures closed the wound. We located telemetered animals by using a handheld receiver and directional antenna to play hot and cold with the signal, despite interference from the rugged terrain and a tiny rainforest frog whose piercing call sounded exactly like the transmitter signal's *tink-tink-tink*.

Manuel Santana, a Salvadoran wildlife biologist, rounded out our team of snake monitors. For weeks we daily checked a pregnant terciopelo basking on a tree-fall, once witnessing her drinking rain droplets off her own scales. Another big female shed her skin while crawling rectilinearly over wet leaves, transforming within minutes from dull and mud-smudged to lustrous velvet. Terciopelos were sometimes diurnally active, but the larger bushmaster, as Manuel, Dave, and I discovered by keeping both species under round-the-clock surveillance, is truly a creature of habit. As night fell and patterns on the forest floor became indiscernible, up went the chiseled head in an alert ambush posture that was maintained until shortly before dawn; just as a filmy gray light settled on the understory, down that head went into a resting coil—night after night, day after day, for up to sixty-seven days at a site. My strongest memories are of phenomenal camouflage. One morning I saw an experienced local woodsman walk by a big bushmaster coiled in plain view at trailside, and another day, after a group of us had talked and taken photographs around a six-foot terciopelo for several minutes, someone remarked, "What about the other snake?" And there he was, a smaller male, completely in the open, lying against her.

La Selva never yielded the sustained observations of individual pitvipers that I craved, but since Dave proved an enthusiastic collaborator, we commenced long-term joint research on black-tailed rattlesnakes in 1985. We suspected that the easygoing, three- to four-foot-long rattler species and its fairly open habitat were well suited for behavioral studies, and Dave's home in Arizona's Chiricahua Mountains could serve as our base. Our two-mile-long field site was drained by Silver Creek and bordered on the north by the slanted, knobby crest of Limestone Mountain; to the south rose Silver Peak and the main range of the Chiricahuas. Massive stumps testified to historically forested slopes, now clothed in pinyon-juniper woodland and low thickets of white-thorn acacia, agaves, and several kinds of cacti. Sycamores and cotton-woods formed shady galleries along the stream. White-throated woodrats, rock squirrels, cliff chipmunks, and desert cottontails provided prey for black-tails, and we recorded coyotes, bobcats, black bears, mountain lions, ringtails, coatis, and gray foxes as potential predators at the site.[28]

Dave and I watched the Silver Creek Canyon blacktails for fifteen years. He drove down from Tucson at least monthly to monitor telemetered snakes, and I'd join him for several weeks during the summer monsoon season. We'd get out shortly after dawn, walking the rocky slopes and locating animals before they took shelter from the midday heat, each of us carrying telemetry gear, capture equipment in case new snakes were found, and plenty of water. Our daily routine was to walk toward the sun, so as not to cast shadows and disturb the snakes, moving low and slowly to within a few yards for observations and photographs. The blacktails often seemed aware of us but typically paused only briefly, rarely rattling, and resumed activity. Once I crawled under a juniper, frustrated by a confusing signal from superfemale 21; after several minutes of scrutinizing litter for a glimpse of scales, I sat up to think about things—and spotted her inches from my face. Coiled on a yard-high branch, she didn't so much as flinch while I edged out of strike range, thankful we'd never traumatized her during capture.

As hunters, blacktails proved to be pretty standard vipers. They waited next to logs, woodrat nest entrances, and rabbit trails for hours or even days on end, head drawn back and ready to strike, occasionally successful but usually not—we estimated that each snake ate as few as three to five meals a year. Woodrats were staple prey, but female 12 seemingly specialized on squir-rels by coiling against tree trunks, with her head pointed up. After swallow-ing prey, the rattlers often crawled up to several dozen yards to a crevice or other shelter, where, safe from predators, they protruded a meal-laden body loop into sunlight to aid digestion. Late one fall, however, female 8 ate what I suspected was a desert cottontail, so big she could scarcely pull up under the nearest shrub, and although I feared she couldn't process her meal because of the chilly nights, after nine days of basking the lump disappeared and she made it into hibernation. We failed to learn much about the biology of young rattlers, though we found one about a month old that had swallowed a brush mouse slightly exceeding its own weight.

Like most vipers, blacktails don't gain enough energy each season to bear young annually, so well-fed, receptive females are always in short supply. By late March both sexes emerge from hibernation high on the slopes and move across the ravines to hunt, except females who mated the previous summer—these will seek out nearby gestation sites and await birth. As summer rains begin, males crawl hundred of yards daily at about a body length a minute, seeking mates. One went between my feet, presumably because I'd straddled a female's unseen chemical trail, and another investigated Dave's pant legs as he sat taking notes; neither snake acted defensive, let alone aggressive. Courtship can last for days. It consists of the male tapping his chin and tongue-flicking along the female's back, then wrapping his tail around hers—at which point she opens her vent or, more often, slaps his tail back with an audible clacking of rattles. After pausing for a minute or two, he resumes courting. If another male approaches, the rival suitors topple each other with their foreparts, a sort of limbless arm wrestling that results in one of them, usually the smaller, leaving. Copulation can last for at least twenty-two hours—and, as we observed, a large male may, even in the midst of mating, fight off an intruder that tries to displace him.

Our most exciting finding has backstory, a long-prevalent view exemplified by this quote from Laurence Klauber's 1956 opus, *Rattlesnakes:* "There is no final evidence that young rattlers stay with their mothers for more than a day or so at most; if they are found together there is no proof the young are more than a few days old or that their propinquity is caused by other than the use of a common refuge."[29] In fact, during the high school massasauga project, I'd found a female and eleven young and noted their cloudy, pre-shed eyes, not realizing this meant she'd been attending them—despite having read a 1942 paper by my Missouri mentor Paul Anderson, who had "considerable evidence that . . . [mother copperheads and timber rattlers] remain with the young for several days."[30] In 1966, Charles Wharton further set the stage, noting that "two adult female cottonmouths were found with thirteen newborn young nearby . . . [an] incident so striking I regard it as guarding behavior. . . . Aggregation prior to and following birth could . . . [repel] predators who could more easily cope with a lone female."[31] Nonetheless, it wasn't until almost forty years later that we published decisive evidence for widespread parental care in pitvipers.

Simplistic notions of snake sociality went out the window in 1995 with a phone call that began, "You're not going to believe this!" Dave had found superfemale 21 lying outside her gestation site, an abandoned rock squirrel burrow, with six babies about three days old, in pre-shed condition. For another week they basked together, and if Dave got close, the neonates fled to safety, with 21 backing in after them. On the tenth day each baby shed in front of the obviously attentive mom; next morning there were six entwined translucent skins, the young had disappeared, and she was forty yards away at a woodrat nest, hunting for her first meal since the previous fall. Some other females, we subsequently learned, never left the nest while babies basked, whereas one advanced on humans and dragged an errant youngster back into their hole. We later showed that parental care characterizes most pitvipers, and in my public talks their maternal behavior has proved especially useful for inspiring empathy for snakes.

Human phobias are deeply rooted but also impressively variable—spiders and heights unnerve me, whereas Philippe Petit, the "Man on a Wire" between the World Trade Center towers, is terrified of "snakes, anything with too many feet or not enough feet!"[32]—which suggests that we might come to better understand our fears and transcend millions of years of snake-primate conflict. In 2011, 254 aficionados gathered at a rattlesnake conference in Tucson, a level of enthusiasm inconceivable when I was a graduate student; among the presentations were field observations of substitute parenting by Arizona black rattlesnakes and experiments revealing that rattlers excel at single-trial learning compared to other pitvipers.[33] I came away wondering what new surprises await us and ever more convinced that research plays a central role in valuing nature, a proposition we'll explore further in Part Three.

Meanwhile, life winds on like a serpent, seeking shelter, food, and sex, never knowing what lies ahead. Explorer John Cadle found an Andean short-tailed snake that froze to death while swallowing a lizard[34]—not usually risky behavior, but at fourteen thousand feet late-afternoon temperatures plummet. Perhaps a nearby boulder cast frigid shadows or clouds obscured the mountainside; for whatever reason, heat dissipated rapidly, and with one last, laborious jaw excursion the rear-fanged killer itself slipped into stillness. As for the blacktails, some we'd watched for years were healthy one day, bones picked clean by ants the next; a few older snakes declined for months, rarely moving, then abruptly drew flies. On a happier note, twelve years after we first met our superfemale, an email from Dave read, "No signal from 21, so the radio's failed and we're done tracking." For all of death's inevitability, I'm thankful for having only pleasant memories of my favorite snake—swallowing cottontails and rock squirrels, mating under a century plant with giant male 26, and lounging in the sun with her young.

PART THREE

Pretty in Sunlight

ELEVEN

Friends

THIS BOOK IS ABOUT STUDYING NATURE, incorporating one's findings into broader biological and societal concerns, and reaping the emotional rewards of those activities. Doing natural history involves people—as I'll show later, observing and recording are primal aspects of *human* natural history—and however much solitude beckons, we're no more truly separate from others than from our surroundings. Friends with whom we've shared failures and triumphs loom in our hearts, and perhaps they're all the more precious to vagabonds like me, for whom so few have spanned life's full arc. The mundane bonds of friendship, like those of family, weave together everything from laughter and tears to memories and dreams of the future, yet too often friends are taken for granted until a good one is lost. Then they gain bittersweet focus; their value feels inestimably high and their loss insurmountable. Friendships, we learn, require nourishment, but they repay the effort manyfold, even after death intervenes.

I first met Ben Dial at an amateur naturalists' gathering in 1962, as he finished high school in Dallas and I began my senior year in Fort Worth. We were pals from those early days in the John K. Strecker Herpetological Society (named for a pioneer Texas field biologist), through lots of thick and a fair amount of thin, both of us enthralled by animals and music. In 1998 Ben learned that his thyroid cancer, facilitated by years of immunosuppressant drugs, had metastasized, and he died within three months, accompanied on his last day by another friend, a hospice nurse, and me. His legacy is a respectable body of research and thousands of inspired students, as well as an example for friends of joy in the face of physical hardships. He often said the decade following his heart transplant was his best, that after the operation he "always took time to smell the flowers." Now, having revisited some of Ben's favorite haunts and scattered his ashes, I see more clearly how passion for life's diversity fueled his happiness.

Ben and I were teenage science nerds, with the then-requisite black-rimmed glasses and burr haircuts, and our interests converged where eastern North American forests give way to prairie and the arid Southwest. We memorized field notes in the Wrights' *Handbook of Frogs* and *Handbook of Snakes,* quizzed each other from Conant's *Field Guide to Reptiles and Amphibians,* and took off searching for herps whenever time and money allowed.[1] We were boundlessly happy exploring limestone bluffs and wooded ravines on the Edwards Plateau, and more so the "mountain islands and desert seas," as Fred Gehlbach's book on the U.S.-Mexico borderlands called the region from west of the Pecos River to the Pacific Ocean.[2] As youngsters, the Texas Hill Country and Trans-Pecos were our havens from big-city class-rooms, and we rejoiced in everything from lichen-covered rocks and the dagger leaves of century plants to the smell of damp land after a thunderstorm and breakfast in small-town cafes.

In those early years we shared an almost mystical fascination with certain species whose unusual attributes intrigued us. Texas alligator lizards, for example, are peculiarly serpentine with flattened triangular heads, forked tongues, elongate bodies, short legs, and long, prehensile tails; they are survivors of an ancient adaptive radiation that includes Gila monsters, Komodo monitors, and snakes, as well as giant marine mosasaurs and other fossil reptiles. Denizens of moist rocky habitats, rarely basking in exposed situations, alligator lizards proved difficult to find and at times hard to capture. We'd spend countless hours searching, only to have one abruptly materialize and just as quickly disappear under an immense boulder. Another would slither out of reach into a tangle of vines and dead leaves. Occasionally we'd succeed and, squirming lizard in hand, remark on its white and copper-brown bands, stare into the greenish eyes, and laugh as powerful jaws clamped down on our fingers.[3]

Although Ben and I later loved teaching, we were mediocre students, too easily distracted by newfound independence, and our professional paths were indirect. He attended Texas A&M University in bouts, interrupted by California surfing and work as a Houston zookeeper, while I finished my degree and spent three years in the army. We'd each published several papers, including one together on field observations of parental care by alligator lizards,[4] and were admitted to graduate programs despite miserable undergraduate records. His A&M master's thesis was on geographic variation in banded geckos, and his dissertation at North Texas State University was on life history evolution in those lizards.

Ben was only modestly ambitious in the usual academic sense, preferring to see new creatures and places or have dinner with friends than to squeeze in more hours of lab time. Despite a heavy teaching load, he nonetheless authored some twenty herpetological titles, the first at age seventeen; in a *Natural History* magazine essay Harvard's Stephen Jay Gould praised a paper Ben had published in *Science* based on his doctoral dissertation.[5] Thereafter he focused on metabolic aspects of unusual behavior, such as aerobics of the aptly named "hatching frenzy" of baby sea turtles during their perilous dash from nest to ocean,[6] and he was the first to measure the energy costs of limbless locomotion in a reptile, the Baja California worm-lizard.[7] He was a tireless advocate for the public understanding of research, winning over many a luncheon group with lectures about how studying organisms, preferably in nature, helps solve environmental and human health problems.

My friend taught briefly at A&M, then moved to Chapman University, a community-oriented private school in California's Orange County, where he received numerous awards and was promoted to professor shortly before his death. He was a famously tough taskmaster, yet his student evaluations were the best I've ever seen for anyone, anywhere. Ben wanted things done right and excelled at making young people think hard about things they'd either never contemplated or grown up taking for granted. He'd set jars of charcoal, salt, and water next to an inverted box on the lectern, then explain that life is nothing more than chemistry, completely comprehensible in terms of basic physical principles. After a few moments, while his class wondered what the hell was going on, he'd lift the box and expose a lively desert iguana! The future of biodiversity depends on how much we appreciate other organisms, and I hope that many Chapman alumni, now influential lawyers, bankers, and so forth, haven't forgotten the lesson of that beautiful lizard.

The most insightful and lasting aspects of Ben's scientific work, as well as his success as a teacher, stemmed from a love of seeing, pondering, and talking about nature. That perspective, an extension of his childhood enthusiasm for the outdoors, also inspired an interest in small, nocturnal lizards. By the late nineteenth century two U.S. species of *Coleonyx* had been recognized: the western banded gecko (*C. variegatus*) of the Mohave and Sonoran Deserts and the Texas banded gecko (*C. brevis*) of the Chihuahuan Desert and adjacent areas. Then, in 1958, William Davis and James Dixon christened the reticulate banded gecko (*C. reticulatus*) based on one they had inadvertently caught in a mousetrap, just east of Big Bend National Park. For more than a decade the new species was known only from that first specimen, and some suspected it was no more than a bizarre variant of the Texas banded gecko.[8] Ben had seen *C. brevis* on our teenage field trips and learned of *C. reticulatus* in Dixon's herpetology course at A&M.

Most of the more than one thousand species of geckos, like the familiar foot-long, orange-and-blue Asian Tokays, have spectacle eye covers and toes armed with thousands of microscopic, hairlike gripping structures. Superb climbers, they're prone to herky-jerky spurts, and as a child in the Philippines I delighted in watching a small brown species sprint upside-down across the bedroom ceiling. Banded geckos, however, are generally terrestrial, with closable eyelids and more typical feet, delicate yellow and brown markings, and short, fat tails. A lunker male might break five inches in total length. They move with deliberate, catlike grace and twitch their tails when stalking insects. Besides the U.S. species, there are three others in the neotropics, and their Old World relatives include leopard geckos, popular as pets. In 1968 a fourth desert species, Switak's banded gecko (*C. switaki*), was discovered in Mexico and later found in southern California as well.

However routine in some ways, banded geckos were also puzzling: Why did the two large desert species have tiny tubercles scattered among their granular scales, like tropical *Coleonyx* but not the widespread western and Texas banded geckos? First Ben experimentally demonstrated that reticulate banded geckos are more sensitive to desiccation than the smaller species, consistent with their small, patchy geography; he still wanted to know, though, what they actually were doing out there in the desert. Poring over museum records, he noticed that the few known specimens were taken on rainy evenings and where roads intersected rocky habitat; thus, further inspired by his captives' prehensile tails and climbing proclivities, Ben got the idea that *C. reticulatus* prefers moist, vertical crevices. Late on a stormy summer night he stopped on a road where a reticulate banded gecko had earlier been found, illuminated the nearby limestone ledges with a lantern, and within minutes spotted one of the reclusive, ghostlike lizards crawling straight up the cliff face![9]

After moving to Chapman, Ben studied Switak's banded geckos in nearby Anza-Borrego Desert State Park. As expected, that species turned out to prefer low temperatures and desiccate easily, like reticulate banded geckos, while resembling in some other attributes the smaller western banded geckos with which it co-occurs. Collaborating with Lee Grismer, an expert on gecko evolution, Ben then used comparative analyses to infer behavioral and physiological changes in *Coleonyx* over geologic time, as banded geckos diversified in the changing environments of western North America. In what was perhaps his most influential paper, they demonstrated that those lizards are descended from tropical stock and that the tiny ranges of the two rare species reflect widespread past distributions, when moister climates characterized the arid Southwest.[10]

Although geckos fascinated Ben, another reptile best symbolized for him the emotional rewards of fieldwork. By the early 1960s less than a dozen Davis Mountain kingsnakes had been recorded since the species' 1901 discovery in an isolated range north of Big Bend National Park. As high school herp nerds keen to make important finds ourselves, we were familiar with the records of every known *Lampropeltis alterna*—two were discovered sheltered among rocks, for example, one of them eating a crevice spiny lizard. At the Fort Worth Zoo we admired the most recently collected animal's black-and-tan bands, marveled at its large eyes and flattened head, all so different from the black-and-yellow speckled kingsnakes we'd caught locally. Moreover, in 1951 *L. blairi* had been described from the eastern edge of the Trans-Pecos region, with dark bands widely split by crimson-orange and obviously related to *L. alterna*. A black-and-white photograph in *Handbook of Snakes* of the second known Blair's kingsnake was spectacular,[11] and that species likewise had since been seen only a handful of times.

Just before heading off to college Ben thought he'd finally collected a Davis Mountain kingsnake, a roadkill that on closer inspection proved to be the equally poorly known Baird's ratsnake, the basis for his second publication, about its distinctive juvenile coloration.[12] One of our Strecker Society friends soon found several Blair's kingsnakes by road-hunting, and within a few years others extended the known ranges of both forms. By the 1980s these beautiful serpents commanded a hundred dollars each in the pet trade, and additional studies demonstrated that Davis Mountain and Blair's kingsnakes are the same species. Those from the eastern edge of the Chihuahuan Desert more likely have red blotches (the *blairi* morph) than individuals from farther west; all of them are now called gray-banded kingsnakes.[13]

Every summer Ben would head for West Texas, searching for what he called simply *alterna,* as well as Trans-Pecos ratsnakes, milksnakes, copperheads, and other creatures that as teenagers we'd regarded as living jewels.

More often he'd encounter western diamond-backed rattlers and other common species, or sometimes just a few toads or a dead black-tailed jackrabbit, but no trip was ever described as a failure. Waiting for dusk, over tacos and chicken-fried steaks, he was energized by conversations with fellow snake hunters about everything from peyote cacti to regional history and politics. More privately, Ben once told me that gray-banded kingsnakes embodied life's complex mystique, from personal disappointments and victories to the as-yet-unknown habits of such difficult-to-observe crevice-dwelling reptiles. "Wouldn't it be incredible," he'd exclaim, "to find out what they're *doing* during the day, when they're *hidden?* That would just be *so cool!*"

Night after night Ben would cruise certain two-lane paved roads into the early hours of the morning, fighting weariness with a cooler of Cokes and scanning for the familiar image of a snake in the headlights. Around midnight he'd pull over, walk off the road to relieve himself, and contemplate whatever answers emerged from the world around him. He'd hear the sounds of myriad insects, of whip-poor-wills and an occasional coyote. Then he'd look up at the stars, drop his gaze to the horizon, and break into a big grin as a light breeze rustled nearby branches and sneaked around his collar. Those familiar shapes on a nearby outcrop were the silhouettes of century plants against an infinite Chihuahuan Desert sky.

Two events profoundly influenced Ben's last two decades and his impact on others. In 1981 he befriended retired businessman Erle Rawlins, and until Erle's death in 1995, at the age of eighty-seven, they set an incredible example for anyone anxious about infirmities and aging. A tall, stout, bespectacled man with thick white hair, Erle was one of the most cheerfully level-headed people I've ever met, interested in everything, and his resolute optimism soon came in handy. The second influential event came late in 1985, when Ben was diagnosed with myocarditis. After a yearlong convalescence at Erle's Dallas home, during which his strength further declined, he was accepted into Stanford Medical Center's transplant program and within a few weeks received a new heart. The next morning my friend was roaming the hospital corridors, jovially intercepting strangers for conversation and walking more easily than he had in months. He was weak from the surgery but ecstatic about a second chance at life.

Ben was an ideal heart-transplant patient because he followed all medical instructions, yet resolutely lived as normally as possible despite impediments.

Oblivious to facial swelling and other side effects that would have threatened the vanity of some of us, he gave riveting lectures about the operation and its impact (one was titled "I Left My Heart in San Francisco"). In the years following surgery, Ben and Erle explored the length of Baja California by car and conquered the difficult road to Batopilas, in Mexico's Barranca del Cobre. They led memorable excursions for Chapman students and alumni to Costa Rica and the Galápagos Islands, and laughed about their six-thousand-mile southern route from Dallas to Orange County by way of Tikal, Guatemala. In my own recollected images Ben is animated, effusively recounting their journeys, and Erle is always smiling.

As much as he loved the Southwest and Mexico, Ben was endlessly nostalgic about California and popular music; a movie about his life would have to be filmed in "La-La Land," with soundtrack by Jackson Browne, the Eagles, and Fleetwood Mac. Although he railed at human onslaughts against nature, whenever I'd say something critical of Orange County traffic and smog, he'd brag about the comeback of endangered brown pelicans at Bolsa Chica Wetlands, near his Huntington Beach condo. Ben found my upcoming move to New York unfathomable because, whatever the lure of exotic travel, he couldn't imagine actually *living* anywhere other than California. And as for rock-and-roll, our fellow herpetologists Janalee Caldwell and Laurie Vitt told me, "He visited us in Topanga Canyon and reminisced about music from the 1960's and 1970's—Ben knew all the lyrics, Laurie played the melodies, and we sang for hours!"

One-sided praise shouldn't obscure my friend's humanity, especially since Ben laughed readily at his own foibles. He was zealously private, phenomenally picky, and downright quirky; a half-century after we first met and fifteen years after his death, I cannot claim to fully understand who he was. Ben was annoyingly dogmatic and judgmental, as well as impossible to pigeonhole politically, since his opinions spanned the spectrum from reactionary to radical. Meals with him often veered into debates—whether or not anyone else really wanted to argue—about evolutionary biology, the environment, medicine, religion, or other aspects of our human predicament. He disliked bullies and stuffed shirts, and he went out of his way to confront people he perceived, rightly or wrongly, to be acting that way. Those things said, though, for most folks, the man was unambiguously likable.

I made several trips to Orange County during Ben's last months, taking him out for meals, discussing the fate of his library and captive snakes, helping him work on one last manuscript, and reminiscing. One shared memory of a trip we'd made together, in the spring of 1979, exemplified our friendship. Ben had just failed his doctoral qualifying exam and flown to California for an escape; I'd recently moved there, mired in a second marriage gone sour. Both of us were emotionally wrung out, and within a day of his arrival we drove down the Central Valley, crossed the Sierra Nevada, and dropped over Tehachapi Pass into the Mohave Desert. We entered a land of shimmering open spaces, mostly creosote bush flats and the occasional stand of Joshua trees, with time to unwind and no specific goals.

We noosed long-nosed leopard lizards and chuckwallas in the mornings, road-hunted for rosy boas and sidewinders by night, and listened to rock-and-roll nonstop. Day after day we were distracted by the endless details of natural history: a desert woodrat's nest of sticks and cactus pads, tucked under a boulder and replete with shotgun shells and old bottle caps; here and there the small, elliptical droppings that when crumbled between two fingers revealed leaves, ants, or grasshoppers as the most recent meals of various lizards, depending on which species had left a particular scat. And little by little, during pancake breakfasts and midday shade breaks, Ben convinced me that I wasn't solely responsible for my wife's unhappiness, while I reminded him of past accomplishments, proof he was cut out for biology. By the time he returned to Texas and I went back to Berkeley, our respective difficulties seemed smaller, more manageable, and we both moved on with life.

With so much past history, I wasn't surprised when Ben abruptly became solemn during one of my last visits to Huntington Beach, said he needed an awkward favor, and wondered if I'd take charge of his remains. Of course I agreed, and we had a good laugh resolving the "big dilemma"—he wanted to be cremated, had thought of at least two places for his ashes, and could I help decide between them? "Hell, yeah," I said. "I'll scatter them all over the place!" After much discussion we agreed on four sites: Erle's grave; into the Pacific, from the beach behind a Big Sur restaurant called Nepenthe; anywhere in gray-banded kingsnake country; and on one of southern Arizona's "Sky Island" mountain ranges.

I'd planned to visit Ben on Saturday, November 7, but he was obviously weakening when we talked by phone on Monday, so I changed flights and got to his home at noon on the 5th. The hospice nurse explained that Ben could no longer get up, had ceased most bodily functions, and was within days if not hours of death. His eyes shifted only slightly when I sat next to the bed, and he managed only a hoarsely whispered "Hi Harry." Sue Lamoreux, whom he'd befriended on a Chapman alumni trip to Costa Rica and who'd virtually adopted him during these final weeks, arrived shortly, and we alternated chatting in the living room and checking on him. The nurse suggested I try conversation, for my own sake and because sometimes, she said, dying people are still listening when they can no longer speak.

So I told Ben I'd miss him, that he'd soon be more comfortable; I barely choked out that he'd been an inspiring teacher and fine biologist, a great friend. He neither moved nor spoke again, and by midafternoon his breathing was slow and shallow. At one point I left for an hour or so, drove to Bolsa Chica Wetlands, and took comfort in the brown pelicans flying over his favorite local spot. Ben died at about five that afternoon while Sue and I were out for a snack, as if in one last stubborn act of privacy he'd waited until we left to pass on. Within a month she shipped me his ashes in a plastic box, bagged in blue velvet.

Ben left me this list of "Some of my favorites" on a computer disk: "I love the music of Buffalo Springfield, the Byrds, Neil Young, Ten Thousand Maniacs, Eric Clapton, and Don Henley; the paintings of Georgia O'Keeffe and Vincent van Gogh; and the words of Robinson Jeffers, David Quammen, John McPhee, and Barry Lopez. I love soapberries collected in the fall and back-lit, really weathered wood, green-glass insulators that are old enough to have a patina, the smell of wet creosote bushes, the colors of a madrone, the antics of desert iguanas, everything about gray-banded kingsnakes, the firepit at Nepenthe, the cathedral at Chartres, the zócalo at San Miguel de Allende, the Study Butte store in Big Bend, and the desert, any desert." Ben loved his students and colleagues, too, the more so for their support during his illnesses. After his death Chapman established an undergraduate biology prize in honor of him and Erle.

We spread some of Ben's ashes in Big Bend National Park that next summer, in the very heart of *alterna* country. Starting from a midelevation trailhead in the Chisos Mountains, Kelly and I hiked for about an hour and a half, first in fairly open country and then down an ever-narrowing and more deeply shaded canyon. From time to time we crossed a small creek that flowed beside the path and I daydreamed about Trans-Pecos reptiles that Ben especially liked—the Texas alligator lizards that fascinated us both and of course his beloved gray-banded kingsnakes. A few yards beyond one last trail bend we reached a towering cleft called The Window, its rocky frame splashed in late-afternoon sunlight and opening to an unseen cascade.

The stream was clear and cool between dappled cliff walls. As a leaf floated past my ankles and disappeared, I wondered if going over the edge would prove fatal and anchored my toes on the rocky bottom. Insects chirped and water burbled while I stared at the open plastic canister. After a few minutes I tossed its flocculent contents toward the Window and for an instant they hung like some magically suspended clot; then a breeze swept down the ravine, and perhaps I only imagined the faint *smack* as thousands of glinting flakes and bone chips burst outward, as if an expanding tan nebula had morphed into cascading fireworks against immaculate blue sky. Momentarily stunned, with my shoulders sagging, I began sobbing into my hands. Walking out, we spotted a Great Plains skink basking on a rock, and that night, over Mexican food in nearby Terlingua, we tilted cold beers in honor of Ben and the glories of Trans-Pecos Texas. Long may they mingle.

A week later we set out with Dave Hardy for Silver Peak, on a high flank of the Chiricahua Mountains in southeastern Arizona, carrying another pint of Ben. The three of us hit the Forest Service hiking trail an hour after dawn, ascending from 5,400 feet to the summit at 7,975 feet in about four hours. We walked in leisurely single file, mostly alone in our thoughts, and from time to time Dave and I teamed up with Kelly to catch lizards for her research. Storm clouds were gathering as we reached the end of the path, on the northern edge of Ben's and my favorite part of the world. The Peloncillos were visible across the San Simon Valley in New Mexico, and to the southeast, beyond that long line of bald granite mountains, we made out the darker, more heavily forested Animas Peak, separated by a shallow pass from the Sierra San Luis in Mexico. There was little talk as we ate crackers, cheese, and apples, and after a few minutes I climbed onto the highest rock.

Remembering the Big Bend experience and unsure about what lay ahead, I flung Ben's ashes steeply upward, imagining they would be blown out over the forested slopes. There were, however, no opportune gusts on that overcast summer day, and his remains fell downward, exploding like dirt clods on the rubble. Immediately a movement below drew my attention to a couple of Yarrow's spiny lizards, covered in gray dust with their heads cocked up as if in disbelief. I brushed ashes off my arms and remembered out loud that our friend's many "favorite" creatures included Sonoran mountain kingsnakes and banded rock rattlesnakes, both predators on spiny lizards and not uncommon at these higher elevations. As we started back down the trail I smiled, realizing that eventually, by way of powder-coated prey, his physical presence might reside in those very snakes.

A year and a half after Ben's death, on my way home from a herpetological meeting in Baja California, I detoured to the Bolsa Chica Wetlands, an hour or two by rental car south of Los Angeles. At the end of a dilapidated boardwalk that spanned the cordgrass and pickleweed marsh, the Pacific Ocean pounding the beach only a few hundred yards behind me, I grimly noticed an inland horizon dotted with oil rigs and, farther back through the dirty haze, a solid line of dull cream and pastel condos. Those wetlands had indeed been "reclaimed," but how, I wondered, did Ben manage so much optimism in this microcosm of overconsumption and environmental degradation? And what was I to make of his life and death? On that day the migratory brown pelicans were elsewhere, replaced by black skimmers and Forster's terns as the most obvious wildlife. My mood was an idle, aching need for perspective, and the birds' knife-winged aerobatics brought to mind miniature Jurassic pterodactyls.

As I turned to leave, several pigeons flew from under the boardwalk. A Hispanic man strolled toward me, his son of about nine on one side and a younger daughter on the other. Suddenly the boy dropped to his hands and knees, put his face right to the slats so he could peer between them, and exclaimed, "Look, Dad, there's eggs, there's *eggs* in this nest!" I heard the joy of discovery in that young voice, devoid of cynicism. The father was attentive to his kids, the little girl grinned as she bent down to look too, and they were still talking about the pigeon eggs as I walked on toward the parking lot. From our talks during his last weeks I knew Ben was at peace, that he wanted us to celebrate collective good fortunes rather than mourn his passing. And with those smiling children fresh in mind, I remembered the final words of my friend's farewell, left on his computer: "Here's my advice for young people, borrowed from Joseph Campbell's *The Power of Myth*. If you are after happiness, it works: 'Follow your bliss.'"[14]

Ben's life and our friendship underscore the questions with which I'm grappling here. How do research and teaching play into conservation, and in particular how do they affect the values we place on biological diversity? What do "wild" and "natural" signify, and how do they influence our search for meaning and happiness? The practice of natural history has been my bliss too, and these closing chapters explore how that happens.

TWELVE

Loose Ends

EARLIER I MARSHALED EVIDENCE THAT our dislike for snakes has roots in ancient predator-prey relationships. The advent of science surely didn't eliminate such prejudice. In 1758, Carl Linnaeus, the Swede who first assigned named species to larger categories called genera, famously maligned amphibians and reptiles in *Systema Naturae* as "foul and loathsome animals, distinguished by a heart with single ventricle and single auricle, doubtful lungs, and double penis. Most are abhorrent because of cold body, pale color, cartilaginous skeleton, filthy skin, fierce aspect, calculating eye, offensive smell, harsh voice, squalid habitation, and terrible venom; and so their Creator has not exerted his powers to make many of them."[1] Thanks to 250-odd years of vertebrate biology, we now know that most of those claims are false, although male lizards and snakes do have two sex organs.

Charles Darwin got off to a better start, reporting in 1839 from his voyage on the H.M.S. *Beagle* that the Patagonian lancehead pitviper's tail "is terminated by a point, which is very slightly enlarged; and as the animal glided along, it constantly vibrated the last inch; and this part striking against the dry grass and brushwood, produced a rattling noise which could be distinctly heard at the distance of six feet. As often as the animal was irritated or surprised, its tail was shaken; and the vibration was extremely rapid. . . . [It] has in some respects the structure of a *Vipera* [Eurasian viper] and the habits of a *Crotalus* [rattlesnake]; the noise, however, being produced by a simpler device."[2] Darwin had presciently implied that the rattle evolved from a sound-making structure present in ancestral pitvipers and lacking in their close relatives,[3] but then he sank back into widespread prejudice, declaring that the lancehead's "face was hideous and fierce; the pupil consisting of a vertical slit in a mottled and coppery iris; the jaws were broad at the base, and the nose terminated in a triangular projection. I do not think I ever saw any thing more ugly."

Two years later, James De Kay's *Zoology of New York* was more accurate than Linnaeus and more complimentary than Darwin: "So general is the repugnance to reptiles that their study has been overlooked, and they have been usually considered as beings which it is not only necessary but meritorious to destroy. A part of this vulgar prejudice is derived from education, and perhaps some of it originates from the fact that several of them are furnished with venomous fangs, capable of causing intolerable sufferings and death. To the naturalist and physiologist, however, those who study nature's modifications of form and structure, they present some of the most interesting objects of contemplation. Their utility, either in diminishing noxious animals, or in furnishing food to others, has been lost sight of; and because they were cold to the touch, with a naked slimy skin without hair or feathers, they have been considered as loathsome and hideous, although their structure displays as much of the omnipotence and care of the Creator as can be seen in those which are considered to be the most gorgeous and animated of his beings."[4]

By the late twentieth century we'd learned a lot more about serpents, so it's surprising to find a modern equivalent of Linnaeus's ignorant sentiments coming from Alexander Skutch, widely acclaimed for studies of Costa Rican birds. Skutch despised predators, condemned snakes with special vengeance, and among raptors praised only the laughing falcon, a coralsnake-eater. In earlier writings he grudgingly allowed for predators in wilderness but preferred a "principle of harmonious association," whereby "every member is compatible with every other and there is mutual exchange of benefits." To justify killing tiger ratsnakes and other nonvenomous species, he labeled them "never really social" and "devoid of parental solicitude," then got really steamed up: "The serpent . . . crams itself with animal life that is often warm and vibrant, to prolong an existence in which we detect no joy and no emotion. It reveals the depth to which evolution can sink when it takes the downward path and strips animals to the irreducible minimum able to perpetuate a predatory life in its naked horror. The contemplation of such an existence has a horrid fascination for the human mind and distresses a sensitive spirit."[5]

Granted, this all gets complicated, and my own exposure to the philosophical problems posed by predators began with a pragmatic rural slant. As a child, I saw Grandpa Gibson resolve snakes versus eggs by dispatching henhouse marauders with a hoe, and raptors soaring over his farm were shot as "chicken hawks." Nonetheless, I began overturning rocks in search of herps,

Natural-born killers: western diamond-backed rattlesnake swallowing an adult male acorn woodpecker at a backyard bird feeder in Portal, Cochise County, Arizona, August 2, 1999. The snake was thirty-one inches long, weighed about ten ounces, and required two hours and forty minutes to swallow its three-ounce prey. (Photo: H. W. Greene)

whereas aspiring ornithologists looked skyward. My interests later converged with theirs because some snakes rob nests, some birds kill snakes—and like some but not all bird lovers, Skutch saw good and evil in those utterly natural dynamics. In one last essay he even argued against conserving tigers and barely stopped short of advocating extirpation of all predators.[6] Of course, some characteristics of birds evolved as defense against enemies, and without predation overpopulation favors disease and starvation, but beyond such factual matters, here is the heart of it: Can we appreciate animals as they are, even if dangerous to us and our livestock, or must we imbue them with human motivations, judge them by our goals and values? Skutch treated favored creatures like exhibit pieces or cultured pearls, existing for his enjoyment, whereas I'd rather watch and wonder what it's like to be a coachwhip or a house wren.

But then again, why *should* humans admire and conserve things that might hurt us? Are anthropocentrism (human-centered perspectives and values) and anthropomorphism (human characteristics attributed to other organisms) really so bad if the fate of the planet is in our hands and biological heritage underlies all life, including our own? And so what if—a real possibility—

however much we yearn for wilderness, humanity no longer accommodates natural-born killers? After all, animal rights advocates admonished George Schaller to provision the carnivores he studied with pre-killed game, so that the "cruel" behavior of wild dogs killing zebras couldn't play out and lion cubs wouldn't starve during a drought.[7] One conservation biologist told me that having experienced Africa's megafauna first hand, she wouldn't want to live anywhere near dangerous animals. Clearly the burden of proof is on those of us who find such creatures worthy of tolerance, let alone love.

In truth, we twenty-first-century humans are saddled with terrible dilemmas, willy-nilly shaping the future yet bereft of consensus over what to save, let alone how to do so. Amid shrinking resources, should we care more about pandas than crocodiles, especially if the latter eat us? How can we yearn for places untrammeled by humans yet also bemoan disconnectedness from nature—can we really *connect* without *trammeling?* In the face of shrinking habitats and climate change, with the inevitable extinction of many species— including perhaps most large, nondomestic animals—what are feasible conservation goals? Not only are concepts like wilderness debated, but it's as if over the last few million years, having gained awareness of our mortal fate, we've pulled away from the very things for which many of us profess such longing. These loose ends occupy my daydreams, the more so as certain journeys—some professional, others personal—have taken me more deeply into nature.

Archival photographs of a dark-skinned, long-haired man with a headband, cheeks streaked with white paint, holding a rifle and crouched in an arid landscape, evoke no more than mild curiosity for those unaware of western North American history. People steeped in the region's lore, however, attach special, sometimes starkly contrasting significance to images of Geronimo and his Apache warriors. Nature's like that, too, insofar as we imbue places and organisms with positive or negative meanings depending on context. The folk taxonomies of indigenous peoples, for example, organize plants and animals as food, medicine, enemies, and spiritual totems,[8] whereas biological classifications emphasize descent with modification. Until recently, though, we lacked an evolutionary take on the question of why humans might find some organisms especially appealing. Put more generally, must something be useful or beautiful to matter? Should Darwin have appreciated that South American pitviper anyway, despite having perceived it as ugly?

People admire birds with dazzling hues, even their stuffed skins in museum cases, whereas less gaudy creatures achieve aesthetic impact through the likes of physical prowess and harmonious habitat relationships—a cheetah chasing an antelope, all blurred spots and flashing hooves on the African savanna, or, for that matter, a Patagonian lancehead's gravel-matching browns and tans, the better with which to ambush lizards and mice. There are indeed cross-cultural trends in how bright color patterns make scarlet kingsnakes and other species especially attractive,[9] and we readily fall for round-headed, warm and furry bears, especially when, as with pandas, their eyes look large and childlike.[10] There must be still more to appreciating nature, though, because some of us like tuatara, coelacanths, and other homely creatures.

At this point we need what humanists call "terms of criticism," and in this regard a 1997 essay by U.S. Forest Service ecologist Ross Kiester lit the way.[11] He noted, first, that although aesthetics are often mentioned as justifying conservation, how and why we appreciate nature has received little attention. Nonetheless, as Ross pointed out, aesthetic properties are durable, enhance enjoyment, and can influence decisions about conservation. Then he explained how Immanuel Kant's 1790 *Critique of Judgment* distinguished between beauty, as a property of individual objects, and the sublime, which transcends them, such that, by providing context for individual organisms, beautiful or not, we might fashion a *biologically sublime* aesthetics. Kant further described a *dynamically sublime* based on power in nature, as with volcanic eruptions, and a *mathematically sublime,* based on sheer numbers. Ross worried that recent controversies about the role of evolution in taxonomy might detract from the aesthetics of biodiversity, whereas I believe that homology and descent with modification— biological heritage and adaptive change—have central roles to play in fostering a love of nature.

Kant's distinction made sense in terms of how, as a teacher, I'd used natural history to change attitudes toward venomous snakes, and shortly after reading Ross's essay about applying the German philosopher's dichotomy to biodiversity, encounters with a different sort of dangerous predator bolstered my conviction that he was on to something. In April of 2000, I'm in Everglades National Park, where a few dozen miles west and a century earlier Mr. Watson, of Peter Matthiessen's wonderful historical novels, was gunned down by his terrified neighbors in a landscape as foreboding as any ever invaded by white settlers.[12] Even now, with a bit of mental squinting, the place feels wild and woolly. This afternoon Kelly, her mom, Sally, and I have visited Flamingo, on Florida's southwest tip, seeking one of the largest living predators. I'd caught juvenile American crocodiles in Panama in the 1970s by eye-shining them from a canoe at night, and, having as a teenager read of this species in Conant's *Field Guide,* I'd long hoped to see one in the United States.[13] Today we've already encountered a twelve-footer, basking with mouth agape and looking as wide as an overturned johnboat.

Two hours later and not far inland, the knobby backs of American alligators break up reflected, low-angled sunlight on Nine Mile Pond. I'm expecting nothing more dramatic than a passing cattle egret and the whine of mosquitoes, maybe a large-mouthed bass breaching the tranquil surface or the snoring jug-a-rum of a bullfrog. Kelly and Sally move off to the left, exploring shoreline, and soon my attention focuses on a gator fifty yards out to the right, then on a much larger one behind it. Through binoculars the big gator looks broad-bodied and oddly short snouted; it swims steadily toward the smaller one. Within a couple of minutes the closer animal veers sharply parallel to shore, and although it's mostly submerged, I discern the telltale rounded head and dull yellow markings, remnants of juvenile coloration—a small adult *Alligator mississippiensis.*

As the larger crocodilian turns I see that the short muzzle was illusory, because most of its head had been under water. *It has a surprisingly long snout—and I've been holding my breath in suspense!* Less than a minute later, they're a body length apart when the big one lunges out of the water, mouth open, its distinctively narrow head and long lower fourth tooth clearly visible: The pursuer is an enormous, olive-green *Crocodylus acutus,* perhaps even bigger than the one we'd just seen at Flamingo! Both animals drop from sight, with the gator soon reappearing twenty feet away and the aggressor surfacing where it went under, facing the smaller one. The croc disappears again, exquisitely tense seconds go by, and water explodes as dark torsos twist out so violently I scarcely perceive scaly limbs and tails in the swirling melee. The gator rapidly swims away as the croc turns back sideways for a couple of minutes, then sculls off toward its original locale. Before the huge creature moves away it looks directly at me, and I have the specific sensation of being *evaluated.*

After the incident at Nine Mile Pond, we drive a few miles and stroll Anhinga Trail. Visitor signage stresses that a gator's brain is walnut-sized and hints that this makes them especially dangerous, as if a large cerebrum might prevent a half-ton reptile from viewing us as prey or territorial intruders!

Dusk approaches and we come upon several motionless gators in water along the boardwalk. With binoculars I scrutinize characteristic external anatomy and wait in vain for so much as a blink, but I don't think about brain size. Their valvular nostrils cycle back and forth between oval openings and tightly closed slits, reminding me of how well suited these powerful animals are to an aquatic lifestyle—they also have elevated, protuberant eyes, narrow crevices for ear openings, and a laterally flattened, muscular tail. A transverse fold at the back of the tongue prevents them from swallowing water while they seize prey, and of course there are those terrible jaws and teeth.

Watching the gators, I never label them stupid. As with other crocodilians, they exhibit visual and vocal social rituals, nest construction, and parental care—like birds, but unlike most turtles and lizards. Indeed, the twenty-three species of crocodiles, alligators, caimans, and gharials, plus more than nine thousand living bird species, derive from a Mesozoic archosaur radiation that included other dinosaurs, winged pterosaurs, and diverse extinct crocodilians.[14] Today the alligators look lazy, but I know their behavioral repertoire includes a high walk, a gallop, and leaping straight up out of water; I know how those jaws that can crush a deer also gently transport their babies from nest to the water's edge. I've seen the Bronx Zoo's captive crocodilians respond to training with the alacrity of my dog Riley, and all this reminds me that we've been mattering to each other for a long time.[15] Almost two million years ago an aptly named Nile crocodile relative, *Crocodylus anthrophagus,* consumed early hominids,[16] while less than two thousand years ago, two Pacific island species, distantly related to extant crocodiles, went extinct likely because of human depredation.[17]

Natural history provides facts about organisms, including the ones Linnaeus and Darwin found so loathsome, and thereby expands our background for using and appreciating nature.[18] From the standpoint of ecology and aesthetics, of appreciating giant reptiles in a broader context, no other vertebrates preempt the crocodilian role of giant freshwater meat-eater. In terms of evolutionary history, crocodilians are the closest living relatives of birds and our best surviving icons for a world once ruled by ponderous herbivores and huge flesh-eating reptiles, one that predated the rise of mammals. Because someone had discovered these things and I'd learned them, that big American crocodile casting an eye my way was the epitome of a biologically sublime experience, one that otherwise would have been less compelling. Later, I did wonder what was going on in *its* walnut-size brain, thought to myself *probably quite a lot,* and felt lucky to stand by Nine Mile Pond at such a provocative moment.

Gordon Burghardt, my Ph.D. advisor, has pursued diverse research over his forty-year career, ranging from serpent feeding behavior to behavioral development in hand-reared black bears and free-living green iguanas. In teaching and writings he's often stressed Jacob von Uexkühl's *Umwelt*—literally, "surrounding world," but defined by von Uexkühl as that perceived by particular animals—and two projects sharpened Gordon's parallel concerns for their inner worlds. Together we showed that hatchling eastern hog-nosed snakes responded to danger by body-flattening, hissing, and striking, followed by death-feigning, in which they writhed about, then lay belly up and still. Most importantly, we discovered that these innate behaviors were influenced by experience and context: snakes rapidly switched back and forth between death-feigning and crawling away depending on whether a threat (Gordon or a stuffed owl) faced them or faced away.[19] About that same time, he began studying a two-headed black ratsnake, nicknamed "IM" for instinct and mind. IM's heads often struggled over food, and dominance shifted repeatedly over the years; one head consistently preferred smaller prey, although each consumed roughly the same overall amount of food.[20]

Now let's revisit Niko Tinbergen's four research aims for ethology, the questions that inspired my graduate research: How is behavior controlled, how does it develop within an individual, what's its ecological significance, and how did it evolve? In 1997, Gordon, having earlier spent a sabbatical with Rockefeller University's Donald Griffin—the discoverer of bat echolocation and pioneering advocate for understanding nonhuman minds—raised a fifth aim: What are the *private experiences* of animals?[21] Griffin had claimed that consciousness and awareness are indicated by surprising yet effective solutions to changing, unforeseen, and uncommon problems, and Gordon realized that snakes are especially tantalizing in that light. Although IM's two heads seemed ludicrously irrational, competing despite a shared need to provision their same body, hog-nosed snakes nicely met Don's criteria, behaving as if aware of deceptive and dangerous relations to predators, acting as if they assessed the dynamic appropriateness of alternative defensive tactics.

Griffin might well have regarded our black-tailed rattler that manipulated its ambush site as having a mind, but I'll wager that in 1976, when he published *The Question of Animal Awareness,* even most biologists wouldn't have entertained that possibility.[22] Indeed, Don took heat from psychologists who regarded that book as too fuzzy for serious science, so Gordon was at pains to carefully label his fifth aim for ethology. *Cognition,* he noted, can itself be viewed as behavior, and thus fails to encompass mental phenomena like emotion, intention, consciousness, and awareness, each of which also might be explainable in terms of control, development, ecological role, and evolutionary history. Moreover, he wanted consistency with Tinbergen's aims and minimal conceptual baggage, as would not have been the case, for example, with the term *subjective.* In ethology and psychology, *experience* means the conditions and stimuli presented to organisms, and *private* implies that the mental phenomena to be studied are accessible to the organism from within. The big question, of course, was how we might learn about private experiences.

Granting that brain scanning and other innovations may well open new windows on the minds of animals, Gordon suggested we begin by combining natural history observations with *critical anthropomorphism*—that is, use human perceptions, intuition, and feelings, *our* inner worlds, to forge novel, testable hypotheses about those of other species. By so doing, he steered clear of uncritical caricatures of other creatures as little more than poorly formed humans—what I call stealing their cat-ness, snake-ness, and so forth—but in the spirit of nothing ventured, nothing gained, he also rejected stifling, narrowly defined objectivism. Later, during fieldwork on green anacondas, he and his student Jesús Rivas nicknamed critical anthropomorphism "wearing the snake's shoes."[23] Why, the two asked, are males of that species so small? Imagine lying for hours in the shallows of a tropical slough, among a dozen seven-foot suitors for an eighteen-foot female, entangling your muscular, scaly tail with others competing for her vent. Perform that thought experiment, they said, and a testable hypothesis comes to mind: male anacondas need to be large enough to beat other males, but not so large as to be mistakenly courted as a female.

Another example of critical anthropomorphism comes from Frans de Waal, who has brilliantly demonstrated that human morality is linked to homologous phenomena in other species. A core tenet of psychology has long been never to accept complex explanations if simpler ones suffice—conscious mental events versus hard-wired responses to stimuli, for example—yet, as de Waal noted, this flies in the face of evolutionary parsimony. The *simplest* explanation for similar behavior among close relatives is in fact that similar underlying neurobiological control mechanisms and internal manifestations (Burghardt's private experiences) were present in their common ancestor. Exemplary among Frans's studies of postulated components of morality in nonhuman primates is one in which he and graduate student Sarah Brosnan asked whether monkeys have a sense of fairness. They offered two capuchins different rewards for completing a task within view of each other, and, like the proverbial picture, a video clip of this experiment is worth a thousand words: when Brosnan hands monkey B a cucumber chip after it's seen monkey A given the tastier grape, B flings the cucumber back at her![24]

A sense of fairness may be restricted to some primates, but I'll bet snakes hold special promise for addressing Burghardt's fifth aim, challenging us to go beyond introspection, language, facial expressions, and nonverbal gestures with which we identify mental events in ourselves and near relatives. Concern for private experiences might even transform how we study snakes and portray them in public, as well as thereby enhance their appreciation and conservation. One day in the mid-1980s, when Dave Hardy and I were manually restraining bushmasters and terciopelos, he said, "We don't have to do this." By then I'd pinned hundreds of pitvipers in the lab and field, but now Dave reported zoo workers immobilizing them with tubes, a method that is safer for people and easier on the animals.[25] Soon thereafter, comparative anatomist Alan Savitzky told me of finding fractured viper skulls in museums, presumably from snakes that had been pinned, and, while reviewing antipredator mechanisms in reptiles, I learned that predators mainly attack their necks and heads.[26] Heavy restraint, I concluded, by mimicking a terrifying natural encounter, might therefore be both psychologically and mechanically traumatic to a snake.

If I were a viper, pinning my head would freak me out, make me fight like hell!

Wearing "snake's shoes" also has affected how I interpret their behavior. Henry Fitch was skeptical that blacktails have consistent home ranges over their lifetimes, and said that Kansas copperheads seemed to just keep traveling—but as a teenager I'd watched him restrain these snakes with a wooden ruler, typical for herpetologists of our generations. I began to wonder if by simulating predator encounters we'd provoked our study animals to relocate, thus encouraging abnormally nomadic lifestyles. I had to acknowledge having pinned venomous snakes out of a misguided sense of necessity, but also because I liked picking them up, especially when others admired my skills: manhandling snakes entailed what naive bystanders regarded as charismatic prowess. Now, however, I had to admit snakes could be studied in the field, collected as specimens, and kept captive without our doing that to them. Finally, a friend who actually needs to pin snakes to milk venom for research has been bitten less than once every hundred thousand procedures.[27] For me the deal breakers were realizing that pinning is potentially traumatic and that the appearance of risk and bravery on the part of those doing it, as a matter for bragging, is misleading. Once I'd faced up to those truths I couldn't keep pinning snakes, let alone do it for the sake of showing off.

Fewer than half of rodeo bull riders make it to the eight-second bell, so if you want to show off some honestly fearless bravado, get on a bull!

Then there's the matter of authentically portraying dangerous animals to the public. In one all-too-common scenario, a snake biologist pins a large rattler, raises its finger-pinched head to the television camera, and shows off glistening fangs; in another, Sir David Attenborough, wearing a face shield, provokes a spitting cobra into defensive venom spraying, remarks to viewers that he'd be foolish not to respect the warning, and moves away.[28] Which approach is easier on the animals and better promotes conservation is a no-brainer—when we gratuitously handle snakes in ways that simulate predation, worse yet describe them as insolent and aggressive, that's much more about our inner worlds than their everyday lives. Luckily, by the time Dave Hardy and I began studying blacktails, he'd convinced me to leave my ego at the door and drop the hands-on approach, a shift that paid off big-time. Surely we'd not have had so many rewarding encounters if those snakes had been pinned, and who knows, maybe things would have gone differently under that juniper tree, when superfemale 21, within inches of my face, didn't strike.

Paul Martin's life spanned boyhood bird watching and a scholarly career, capped by daring plans to rewild North America. His is a story of arduous fieldwork and exciting discoveries, of intellectual puzzles and the theory that humans, shortly after our New World arrival, hunted most large mammals to extinction. Paul was a lover of tempting diversions, a visionary time traveler who thought across millennia and continents as easily as most of us locate parked cars—and could anyone else make "chest-deep in giant ground-sloth dung" sound magical? After identifying globe mallow in the thirteen-thousand-year-old manure, this voracious naturalist tried out its leaves and flowers on his own gut! Paul affably confronted critics' proposals that "over-chill" (climate change) or "over-ill" (disease) wiped out the Pleistocene megafauna, and that his "over-kill" theory was culturally insensitive. Along the way, he urged us to regard horses as repatriated natives rather than feral pests and use elephants as ecological surrogates for mammoths, a controversial proposal that jump-started my own preoccupation with the meaning of wild nature.[29]

Paul had advocated replacing extinct North American browsers with Old World proxies in a 1969 *Natural History* essay that, along with two other publications, was the intellectual impetus for what we have come to call Pleistocene rewilding.[30] In 1980, Michael Soulé, who cofounded the Society for Conservation Biology, pointed out that thanks to habitat fragmentation, small populations, and shrinking genetic diversity, the adaptive evolution of large animals was mostly over—for me a stunningly sad revelation.[31] Then in 1982 Dan Janzen and Paul identified jicaro and several other tropical plants as anachronisms, species whose seeds were dispersed by Pleistocene horses, gomphotheres (elephant relatives), and other extinct mammals, which implied there were still roles for those animals to play in modern ecosystems.[32] Nonetheless, it was almost twenty years before Paul's article with David Burney in *Wild Earth* ignited the rewilding controversy.[33] They touted elephants as the ultimate in restoration ecology but initially gained no traction whatsoever with fellow conservation biologists.

I'd been friends with Paul for years, thanks to his Tennessee seminar visit in the 1970s, and my Ph.D. student Josh Donlan was also a fan. Moreover, Josh's masters' research entailed removing exotic rodents and predators from islands, so he was steeped in the conceptual and technical complexities of restoration ecology. During the summer of 2003, bouncing along a Sonoran Desert backroad in my pickup, we decided maybe Martin and Burney weren't nuts after all—sure, bringing lions back to North America was a stretch, given their potential consumption of people, but what about cheetahs and Asian elephants? Were there sound reasons underlying the traditional conservation benchmark of 1492, or might something much older yet more ambitious be relevant?

After the National Center for Ecological Analysis and Synthesis spurned our workshop proposal, we cut half the invitees, garnered some foundation funds, and asked folks to pay for their own travel. The Ladder Ranch, a New Mexico property owned by media mogul–conservationist Ted Turner, agreed to host us, and on a beautiful fall Friday afternoon we began fashioning a manifesto. Our visionaries were Martin, Burney, Soulé, and Earth First! founder Dave Foreman. Grassland ecologists Jane and Carl Bock, mammalogists Joel Berger,

Jim Estes, and Gary Roemer, and paleobiologist Felisa Smith provided level-headed expertise. Turner's ranch manager Steve Dobrot and his endangered species biologist Joe Truett were welcome guests, while a fossil mammoth tooth on the bunkhouse dinner table served as our talisman.

The Ladder Ranch sponsors a Mexican wolf recovery project, so at dusk Steve led everyone out on a canyon rim and let out a howl; I doubt there was a dry eye among us as distant *lobos* responded in kind. Saturday we argued heatedly and good-naturedly, all day and into the night, over justifications, prospects, and objections. The group agreed that "deep rewilding" smacked too much of deep ecology, then sifted through exemplars of a range of related-ness to extant North American species, conservation status, difficulty of restoration, and danger to people: horses and donkeys; camels and llamas; lions, cheetahs, and elephants; and bolson tortoises. By lunch Sunday we'd outlined a manuscript, and even the Bocks, initially critical of the Martin-Burney proposal, signed on. Then Steve and Joe gave us a ranch tour, highlighted by a black-tailed prairie dog town ("Ted likes them," Joe said, "now he's got sixty thousand") and a pictograph we surmised was a bolson tortoise.

Imagine a Serengeti vista on our Great Plains, where millions of bison and dozens of other large mammal species once roamed, and realize we are not conjuring something irrelevant to current reality. Until twelve thousand years ago—only three or four times the age of a bristlecone pine—our megafauna was as rich as Africa's. Besides what little we still have, there were mastodons and mammoths, horses and camels, armadillos the size of Volkswagens and ground sloths bigger than refrigerators—which in turn were eaten by short-faced bears, dire wolves, saber- and scimitar-toothed cats, and larger versions of modern cheetahs and lions.[34] Those creatures are mostly gone now, vanished in a heartbeat of geological time—the last known mammoth lived less than four thousand years ago, about the time the Hittite king Mursuli sacked Babylon—but most plants and other animals with which they coexisted survive today. Some, like Osage orange and devil's claw, persist because we and our livestock disperse their seeds, while others only make sense in a deep-time framework; the pronghorn's phenomenal eyesight and speed, for example, are likely adaptations to predation by extinct chee-tahs. Indeed, our entire biota evolved in the context of a megafauna, complete with mountains of megafaunal poop and megafaunal carcasses. Its time without a megafauna has been minuscule by comparison.

Now, admit that most surviving representatives of what not that long ago was a *global* megafauna are in Africa—a rapidly changing landscape on which people are killing each other over shrinking resources. Almost everywhere, large animals are reduced to fragmentary ranges and low numbers, genetically ill-equipped for long-term evolutionary adjustments. Accordingly, we envisioned Pleistocene rewilding as partially restoring lost ecosystem function and species diversity in North America, but also providing additional populations with which megafauna could adapt to global change. We imagined three stages, the first already under way, since horses, donkeys, and llamas are widespread, and they could be managed and appreciated as ecological participants rather than pests. Stage two would be carefully controlled experiments—hundreds of captive cheetahs and Asian elephants are already in the New World, so why not explore how they'd fare under more naturalistic conditions? Finally, we proposed Pleistocene history parks, thousands of square miles in size, supporting large herbivores and predators, to be managed for ecotourism and bison ranching by Native Americans and other stakeholders.

Reactions to our 2005 *Nature* paper[35] were swift, polarized, and briefly overwhelming. Within days, Josh and I had fielded hundreds of emails and sat for some fifty radio and television interviews. Seventy percent of seven thousand people polled by MSNBC were positive, and the *Economist, New York Times,* and *Journal of Biogeography* offered enthusiastic editorials. Some prominent biologists called our proposal optimistic, and wild horse advocates loved it until I mentioned predators large enough to take the stallions. Negative feedback ranged from polite and thoughtful to hostile and absurd, with one guy threatening to shoot Josh and his elephants and another linking me with an "international Jewish conspiracy." A letter from pseudonymous, and thus cowardly, Joseph Spicatum linked me to "terrorist atrocities like 9-11," called me a "goofball, dipwad, doofus with a scrambled brain," and advised I "stick to playing with lab rats, befriending cockroaches, or collecting dust mites."

As for professional criticism, a couple of mainstream conservation organizations were worried we'd divert funds from their projects already under way—a charge that could be leveled at any new initiative, as opposed to evaluating each on its merits. Critiques from conservation biologists and behavioral ecologists, cited and rebutted in our 2006 *American Naturalist* paper,[36] echoed widespread complaints along the lines of, "Don't you know about rabbits and cane toads in Australia?"—a continent with almost no placental mammals and no native toads, hardly comparable to using Old World cheetahs as proxies for close relatives here less than fifteen thousand years ago. We countered complaints that Pleistocene rewilding candidates aren't identical to what was lost by referring to North American peregrine falcons, whose once-endangered populations were augmented with birds from seven other geographic races, including several from Eurasia, and no controversy.

More disappointingly, rather than address problematic restoration benchmarks and the perilous status of megafauna globally, those critics stooped to hyperbole and falsehoods. They asserted without foundation that we believe "the flora of North America is essentially unchanged since the Pleistocene," then labeled "Pleistocene rewilding... only slightly less sensationalistic [than Jurassic Park]"[37]—but the Jurassic was fifteen *thousand* times as long ago, and, whereas we advocated rewilding with species that were lost recently or their close relatives, the nearest living kin of *Tyrannosaurus* are birds! Those critics added, "Ironically ... the same [issue of] *Nature* documents ... humans killed by lions in Tanzania," but disingenuously ignored that report's prominent subheading: "understanding the timing and distribution of attacks on rural communities will help prevent them."[38] As it happens, ten of the top twelve causes of death in Tanzania are diseases, the other two being human violence and car wrecks; lion attacks don't make the top fifty.[39] Finally, our critics made the puzzlingly erroneous claim that the bolson tortoise can't exemplify Pleistocene rewilding, because it lived in New Mexico during historic times.[40]

In fact, those hundred-pound turtles demonstrate how things might proceed, as well as underscore the most telling objection to Pleistocene rewilding. *Gopherus flavomarginatus,* cousin to the smaller gopher tortoise of Florida and Georgia, was first discovered in 1958, still roaming north-central Mexico's Bolsón de Mapimí, though as fossils demonstrated, until the end of the Pleistocene it was much more widely distributed in the Chihuahuan Desert.[41] Subsistence hunting and habitat change now threaten the remaining populations of those tortoises. When I proposed them among the rewilding exemplars, I expected skeptics to bring up disease, local ecological shifts, and genetic differences between restoration animals and those that had been lost. But as no critics objected to returning the bolsons to the United States, I infer that deep down what really bothered folks was that big cats and elephants, unlike turtles, are dangerous. In other words, NLIMBY: *no lions in my back yard!*[42]

Our first night at the Ladder Ranch, Jane Bock, not yet knowing our agenda, said to me, "You're a herpetologist; we need advice. Ariel Appleton, who owned the Appleton-Whittell Research Ranch near Tucson where we've worked for decades, just died. In the 1970s, she brought a handful of bolson tortoises out of Mexico, and her family wants to do something creative with the resulting twenty-six animals."[43] As my jaw dropped, Joe Truett, sitting next us, chimed in with a grin, "Ted loves turtles." Long story short, Turner funded a strategy conference, health screening, and transport of the Appleton tortoises to one of his New Mexico ranches on which plant communities closely resemble those where the species still lives in Mexico. A single animal died the winter following translocation, but the others have thrived, raising the possibility that bolsons may someday have two widely separated populations with which to face climate change.[44]

So what have we learned? That many conservationists are surprised that mammoths so recently went extinct and North American peregrines were reconstituted from Eurasian stock. That many people are averse to "playing god" with the environment, the more so with dangerous animals—Henry Fitch documented copperheads shifting diets as encroaching woodland forced out grassland prey species, yet was uneasy about restoring prairie on his Kansas study site,[45] while only decades after the extinction of panthers in Florida's Panhandle, outraged locals protested that reintroduced cats would eat their kids.[46] And as philosophical critiques make clear, we still lack broadly justified conservation benchmarks, be they five hundred or fifteen thousand years ago, be they single species or so-called intact ecosystems.[47] For me, Paul Martin's clarion call to save the megafauna still rings true, as does a suspicion that for many folks, embracing life's violent side is especially problematic in terms of caring about nature.[48]

In point of fact, shortly before the Ladder Ranch meeting a rattlesnake reminded me that living with natural-born killers isn't easy, and placing that incident in perspective has involved a lot of miles walked, before and since.

Eleven years old, crammed in a car with four older Boy Scouts and our troop leader, I set out from Arkansas for Kentucky to hike fifty-seven miles in four days. I had no idea what lay ahead, only recently having trudged a personal best of five miles under the watchful eye of my dad in a nearby car. With scoutmaster tutelage, we'd hammered pack frames out of riveted wooden slats, fitting them with padded webbing for shoulder and waist straps. The first day, my new boots were stiff as cardboard, and after twenty miles on the Abraham Lincoln Trail two other boys dragged me blistered and whimpering into our camp near the former president's Knob Creek child-hood cabin. Next afternoon, we arrived at his birthplace, aimed three-fingered salutes at an American flag, and recited his 272-word Gettysburg Address. Then we did the nearby twenty-mile Daniel Boone Trail in two more days. One of my medals from the trip is inscribed, "Scout Harry Greene walked in Lincoln's footsteps," but I was too young to fully reap the rewards of long-distance hiking.

My next backpacking trip was in 1983, traversing montane Costa Rican rainforest between La Selva Biological Station and Braulio Carrillo National Park. In two weeks our team of seven scientists, two trail cutters, and a cook covered distances I could perceive on a map of Central America, our goal to survey biodiversity across an intact elevational transect.[49] Clothes never really dried, and perhaps deafening rain and fetid socks in my tiny tent inspired several bizarre, predawn dreams—once I was a tick on a showering woman's underarm hair, but over and over, just as I contemplated latching onto her breast, water sloshed me off like some acarine Sisyphus. During daylight hours we photographed black-and-yellow harlequin frogs mating beside a stream, made paraffin casts of jaguar and tapir tracks, and otherwise surveyed the flora and fauna. As dusk fell, beat up by the steep muddy slopes and sated on beans and rice, I reflected on dichotomous activity cycles—among frog-eating serpents, we always found three species of parrot snakes and a racer by day, two more of blunt-headed vine snakes and two of cat-eyed snakes only after sunset. As for diurnally adapted primates, when campfire talk turned to what if one's lights failed on a night walk, answers were unanimous: wait for dawn rather than walk and risk snakebite, tripping on a root, or careening off an unseen embankment; wonder what it'd be like to meet a jaguar in the dark.

By 1991 I was with Kelly, and together we marveled at the seven-hundred-year-old ruins of Keet Seel in Arizona's Navajo National Monu-ment—ten miles of downhill switchbacks, a rickety ladder, and leisurely viewing of the cliff dwellings, then up out of the red gorge the next morning. In 1995, just before getting married, we passed four glorious days backpacking the slickrock country of Utah's Paria River Canyon, which empties into the Colorado just upstream from the Grand Canyon. At dawn on day one we dropped into Buckskin Gulch—twelve miles of convoluted, yard-wide pas-sageways, under colossal logs washed in by flash floods, past two live rattle-snakes and the rotten carcass of an owl—and camped on a bench by late afternoon. For the next couple of days, ambling the canyon in blissful solitude, we scrutinized petroglyphs of bighorn rams and scorpions, photographed chuckwallas and leopard lizards, and contemplated the ruins of old Mormon settlements. Nights we talked about bats and planets, listened to the trills of red-spotted toads. By the fourth day our water filters had clogged, but we found a spring and made it out without having to drink from the muddy river, determined to make our lives together.

Visiting Kelly's Peruvian homeland as 1996 ended, we dined on *cuy* (guinea pig) and prowled the Indian markets around Cusco while acclimating to high elevation, then hiked the fifty-five-mile Inca Trail without guide or porters. At one point, as I trudged through the puna grasslands of Warmiwañusca—Dead Woman's Pass, fourteen thousand feet above sea level—an elderly Quechua wearing shorts and sandals, hefting something the size of a small kitchen stove, asked me as he strode by, "Señor, cuántos años tiene usted?" We had fancy gear and by then I was resting every few dozen steps, but my "Cincuenta y uno!" brought a smile to his weathered face. That night, New Year's Eve, over fettuccini Alfredo made with powdered milk and canned crab, we recalled the marsupial treefrog found hopping in a bog by our tent and a hovering giant hummingbird seen earlier, so large its beating wings were visible. On the fourth day we strolled through the Sun Gate, down to Machu Picchu, and marveled at how its stone ruins harmonized with the surrounding cloud forest. Later that afternoon, local teenagers in the steamy public pool in Aguas Calientes regarded this bald, hairy Caucasian with shocked curiosity, as if I were some supernatural *oso!*

Kelly is sharp-eyed, often seeing animals before I do—in Sonora's Sierra San Luis, hiking behind seven male naturalists, she alone spotted a ridge-nosed rattlesnake in trailside leaf litter. Nonetheless, on August 8, 2002, stalking Yarrow's spiny lizards in Arizona's Peloncillos Mountains, she heard explosive rattling, jumped backward, then saw the blacktail and two bleeding punctures on her shin. There followed an hour's drive, during which she mentally isolated the branding iron fire to below her knee, and a little bordertown hospital, where she was stabilized with IV fluids, antivenin, and morphine for a helicopter flight to Tucson. Next came two weeks of hospitalization, including an emergency fasciotomy to relieve swelling, removal of the 70 percent of her anterior tibial compartment muscles that was necrotic, and a prognosis of permanent foot-drop. During two more weeks in a motel, I nightly changed the dressing on her fifteen-inch open surgical wound. Finally, back in Ithaca, we were surprised that some of our friends thought she'd blame the snake or even me, since I studied its species.

After a year of hardcore physical therapy and surgical removal of internal scar tissue, Kelly's feet were working fine and we returned to Paria Canyon. At first my pack shoved down on tender shoulders, a reminder of how long it had been since I'd done this, of my fragility. We were struggling, my load and me, but by the third day the pack was comfortable—one foot in front of the other, mile after mile of boots on rocks, gravel, and occasional quicksand. As my mind looped back and forth, old losses and fears didn't make so much noise or bump each other as often, left room for other things to fall into place. I drifted mentally for hours on end, reflected on time's passage in a cotton-wood-grove campsite where, in 1995, we'd been delighted by orange-headed spiny lizards leaping after low-flying June bugs. As for Kelly, always out in front, she'd run three marathons since the snakebite and was publishing a paper on how the severity of her symptoms might have been ameliorated by better-informed treatment.[50] For her, that hike was a piece of cake.

Both trips to Paria we'd watched without success for California condors, arguably examples of Pleistocene rewilding. Back in the early 1980s, soon after I arrived in Berkeley, a bitter controversy erupted over what to do with the last few individuals of the species. Some environmentalists claimed the birds would irrevocably lose their wilderness spirit if they were subjected to captive breeding, and a Museum of Vertebrate Zoology colleague argued against expensive management of what he viewed as a lost cause; however, a consensus to save condors won. I never saw those big vultures in California, but within a decade tens of millions of dollars were spent and now they fly over several western states—the catch being that $5 million a year are spent provisioning them with cattle carcasses, curing them of lead poisoning, and so forth.[51]

In 2010, my brother Will, our wives, and I celebrated anniversaries by rafting the Colorado. Still searching in vain for condors, we were rewarded by a journey through geological time, from reddish, 240-million-year-old Moenkopi sandstone at the Lee's Ferry put-in to somber, dark, almost two-billion-year-old metasedimentary rocks of the Vishnu Schist, where we left the river. And I couldn't have been more pleased by the pink and tranquil Grand Canyon rattlesnake Kelly spotted next to our tent one afternoon.

For once, however, my emotional highpoint of the trip involved birds. It came near the end of a long walk up Bright Angel Trail to the South Rim, nine miles and 4,200 feet in elevation from where the rafting company dropped us off. A few years earlier I'd walked into and out of the canyon with a much heavier pack, but now I was the oldest and slowest among seventeen hikers. Late that afternoon, pausing for water on the next to last switchback, I was fantasizing about a Native American platform burial, imagining how cool it would be for coyotes to carry my bones toward a setting sun and otherwise be strewn about by scavengers. Then I glanced up at an enormous storm cloud against a perfect blue sky. Soaring overhead, highlighted by the silver-white background, were two black birds, so much larger than turkey vultures that, even without mentally ticking off light wing linings and slotted primaries, I knew they were California condors. "I'm not ready yet," I shouted with upraised fists; then, smiling: "You're worth every penny!"

All those miles walked, with so much joy and always more to learn. Coteaching a field course in Kenya, five years after the second Utah hike, yielded insights at odds with the mindsets of many environmentalists.[52] First, granting that modern peoples are taking terrible tolls on African biodiversity, defining wilderness by absence of human impact on the continent of our origin would require going back to before we were *Homo*—and even that'd be dicey, because surely australopithecines were ecological movers and shakers too. Other large landmasses also are problematic, since in North America, for instance, we've irrigated crops, managed fire, and so forth for thousands of years.[53] Second, in Kenya, our students assessed megafaunal numbers by counting dung, from dainty antelope pellets to the horse apples of zebras and cannonballs of elephants. Lots of poop everywhere, I realized, even in the waterholes, *is* pristine! Finally, ubiquitous death brings its correlates, fear and demonization. We chronicled the changing colors and stench of a giraffe carcass, then watched as a hyena lugged off its grisly head. On foot, I routinely checked wind direction, skirted thickets that might hide a lion, and most tellingly, came to regard the magnificent elephants that charged our vans as neighborhood thugs.

These perceptions were reinforced in 2011, when Kelly and I visited South Africa's Kruger National Park. During twenty miles of off-road hiking we had numerous close encounters with white rhinos and elephants, as well as several species of antelope, rollers and other colorful birds, and a spectacular orange and turquoise flat lizard. I was most inspired, however, by Cape buffalo, which have long impressed me as the crustiest, scariest, meanest, dare I say evilest-looking beasts on the planet. Yet when armed park rangers guided us through a scattered herd, skillfully avoiding the animals' discomfort zones, I came to see how those huge ungulates, under constant threat of predation, survive with every means at *their* disposal—including charging whatever scares them. Horrifying tales of fatal tramplings and gorings notwithstanding, I had to admit, Cape buffalo are just as *unaggressive* as rattlesnakes.

Pondering these loose ends—questions answered and some still nagging—I urge young nature-lovers to seek out three sorts of experiences that have pivotally influenced my worldview. The first would be to go backpacking: to feel responsible in your sore, uncertain bones for food, water, shelter, and unforeseen calamities; better yet, at least once, to carry out feces, as we did in Paria Canyon, holding oneself accountable for even personal waste. Another experience would be to walk among Africa's megafauna, thereby getting a firsthand sense of what we've lost in North America, what we stand to lose soon all over the world, forever. And third, I would encourage novice naturalists to slaughter and eat a large mammal—the vegetarian alternative being to clear a chunk of habitat that houses several rabbit-sized fur-bearers, dispatching every living thing before sundering the ground with a plow and planting fruits, grains, and vegetables. During decades of studying predators I've often wondered what it would be like to face fully the implications of an omnivorous primate lifestyle. But we'll head down that path in the next chapter.

THIRTEEN

Born-Again Predator

OF THOSE TWO FAVORITE CHILDHOOD books mentioned in chapter 3, the influence of *Snakes of the World* is obvious, that of the other less so.[1] *Stocky, Boy of West Texas* recounted an orphan herding cattle, bringing venison back to the campfire, and racing "horned frogs" for pocket change—and as a kid I'd caught the rotund little iguana relatives that Grandpa also called "frogs," even built a scaled-down version of Stocky's sod house in our family backyard. More than half a century passed, though, before I experienced other elements of his pioneer life, and by then I'd written a dissertation on behavioral evolution, made a career of studying nature, and begun promoting historical perspectives for ecology and conservation. I'd struggled up and down the Inca Trail, carried a heavy pack into the Barranca del Cobre and barely made it out; I'd radio-tracked black-tailed rattlesnakes and bushmasters, even grabbed that huge green anaconda by its tail in a swamp. My path, safe to say, has been fairly wild by some standards, woefully tame by others.

At some point during those wanderings Robinson Jeffers's poem "The Answer" began tugging on my psyche: "Organic wholeness, the wholeness of life and things," he wrote, "divine beauty of the universe. Love that, not man apart from that, or else you will share man's pitiful confusions, drown in despair when his days darken."[2] Of course, we all carry the legacies of Darwin's "descent with modification," but environmentalists often behave as if engaging Jeffers's "wholeness" is more an idyllic pastime than an imperative—Sierra Club founder John Muir even claimed never to have seen a drop of blood in all his hiking, foreshadowing contemporary views of a benign great outdoors.[3] It's as if by denying mortality we might inhabit a peaceable kingdom, and recent circumstances let me ponder this very grownup dilemma while revisiting youthful fantasies. What does it really mean, I'm asking, to be part of nature?

Before moving from Berkeley to Cornell, I'd acquired an old Winchester, like the rifle Bronx Zoo explorer William Beebe might have used to kill a Burmese bandit, the same model with which *A Sand County Almanac*'s Aldo Leopold had shot wild turkeys in the Southwest and poet Gary Snyder would finish off hunter-injured game in his beloved Sierra Nevada.[4] So much for clichés about nature lovers, let alone Ivy League professors. As 2009 wound down I headed back to the Lone Star State with my old carbine, to herd longhorns and hunt white-tailed deer on David Hillis's Double Helix Ranch. My friend was squabbling with his university's president over the football coach's two-million-dollar raise, and I'd been offended by a colleague who equated learning kinds of organisms with memorizing chemistry's periodic table of elements. It was high time we set aside those annoyances, or, as David said, "get out and act a bit more like wild primates." As it happened, I'd just read about savanna-dwelling chimpanzees that fashion saplings into spears, with which they kill lesser bushbabies for food.[5]

I was further inspired that fall by teaching a seminar in which students hashed over hunting, zoos, and other dicey topics. Should wilderness, they asked, be defined by absence of human control, thereby *self-willed,* as Earth First! founder Dave Foreman asserts, or is it simply a romantic construction of postcolonization Americans, as claimed by social theorists? Among other viewpoints we reviewed, I came away impressed by philosopher Baird Callicott's emphasis on preserving biodiversity, the very fruits of evolution, and Leopold's well-known "first rule of intelligent tinkering," loosely paraphrased as, "Save as many of the parts as possible."[6] Ornithologist Alexander Skutch, on the other hand, urged saving only harmoniously compatible species, not spending a penny more on tigers and other predators. He was wackily out of touch with ecological reality, as discussed in the preceding chapter, but many self-styled environmentalists also won't accommodate natural-born killers anywhere near their own backyards. In other words, to recast more pointedly the issues raised by Jeffers, do snake-hating bird watchers love nature more than subsistence farmers like my East Texas grandparents, who supplemented their larder with doves and quail? Can spectators, no matter how adoring, be as wild as participants?

David and I first got acquainted when as an undergraduate he wrote me about seeing previously unrecorded behavior in snakes. As youths we'd each worked for Henry Fitch, and I've enjoyed following David's successes, including a prestigious MacArthur Fellowship and election to the National Academy of Sciences. His professional acclaim comes from using molecular genetics and computational analyses to untangle relationships among organisms as diverse as bacteria, frogs, and the HIV virus, but I admire even more David's dedication to teaching, his unabashed love of natural history, and his adventuresome spirit—he'd named a newly discovered Ecuadorian snake *Synophis calamitus* for "the landslide that forced . . . [us] to stop and collect, and the second landslide that shortly thereafter crushed our field vehicle."[7] Now his Austin home is decked out for the holidays, the kids having returned from college, and upon my arrival we catch up on gossip over Ann Mackie Hillis's green chili chicken soup.

Next morning, with David's former Ph.D. students Greg Pauly and Tracy Heath, we buy groceries and purchase hunting licenses. Texans born before 1977 don't require firearms-safety certification, a hilarious wrinkle that in my case is justified thanks to Grandpa, Boy Scout training, and Lyndon Johnson's insistence more than forty years ago that I join the U.S. Army. As we chat Greg catches me up on his studies of toad evolution and mentions a couple of nice coincidences: his first inkling of a profession for folks interested in amphibians and reptiles came from reading a paper by Fitch, and, like David, he'd once written me about snakes. Then Tracy throws out that her undergrad mentor was Boston University professor Chris Schneider, himself among the Berkeley students and postdocs who'd presented me with the Winchester. In the meantime, she's become a computational evolutionary biologist and developed an interest in cooking personally harvested meat.

We stop at the Hillis's nearby weekend hideaway to test our copper ammo, adopted because conventional lead bullets fragment badly, poisoning vultures and other scavengers as well as human consumers of wild game. While David sets up a target, Tracy sees an injured buck in the brush, which Greg summarily shoots but finds too damaged by infection to save for meat. The deer was flailing when first spotted, and we're taken aback by its nearly severed right front leg and shrunken muscles, leading us to wonder whether the horrible wound resulted from some hunter's bad behavior. After knocking off a few satisfactory shots at a paper bull's-eye, we're off to Cooper's BBQ in Llano for dinner and then on to the Double Helix. We'll be joining up with conservation geneticist Cody Edwards from George Mason University, a Hillis friend who'd grown up in rural North Texas.

Our destination is near Pontotoc (a Chickasaw word for "land of hanging grapes") in the Hill Country, famous for picturesque old German towns, wildflower extravaganzas, and deer hunting. David's ranch is rolling oak savanna, dissected by ravines and dotted with Precambrian metamorphic outcrops. Several rivers drain the surrounding Edwards Plateau and flow on through the Gulf Coastal Plain, favoring moisture-loving species in this otherwise semiarid environment. The region is among the most biologically diverse in North America. About sixty-five species of amphibians and reptiles inhabit the ranch, spanning gray treefrogs and eastern box turtles to the more dry-adapted Couch's spadefoot toads and western diamond-backed rattlesnakes. The Double Helix sits on an overlap zone for eastern and western spotted skunks, so its thirty-five species of mammals include a textbook example of how the different breeding seasons of those two species promote genetic divergence.

David and his sons built the ranch's first dwelling, a minimalistic contraption whose dark-green scrap lumber and tin blend nicely with surrounding rock outcrops and foliage. In the Chateau Vert—named by a visiting French Canadian biologist—one room is used for cooking and eating, while another has two sets of bunk beds; together these rooms could be squeezed into an average suburban kitchen. Amenities include propane lantern and three-burner stove, rainwater reservoir, outhouse, and BBQ grill. A nearby discarded appliance, so battered and bullet-riddled I fail to discern its identity, serves as backstop for target practice. Cody's got the wood stove burning when we arrive, and although others come and go, these four will be my core companions for the week.

As if the landscape weren't sufficiently uplifting, things are good socially too: in spite of a half-dozen advanced degrees and our wide age range, no one seems out to prove anything. Cody's a lanky cowboy, David and I are middle-aged mesomorphs, and Greg and Tracy are a bit shorter—he looks like a trim wrestler, and she can't weigh a hundred pounds soaking wet. Tracy and I are newbies, never having hunted big game before, although two years ago she watched David kill a deer. My .30-30 has open sights (its 1894 design ejects spent cartridges upward, making scope mounting awkward), and the tree stands favored here challenge my acrophobic comfort zone, but David cheerfully indulges these limitations, adding that among the commonest causes of death for Texas hunters is falling from their perches. Tracy, who will use one of David's scoped bolt-action Rugers—longer shooting rifles than mine—seems up for anything.

Greg wakes us at 5:15 A.M. for coffee and cereal bars, so everyone can hike to well-separated hunting areas before first shooting light at 7. Dawn is breaking as I settle into the brush-covered Chateau ground stand, around the upper end of a ravine from the cabin, and peer downslope toward a game feeder that twice daily slings out corn. Out of nowhere I remember Yale biologist Evelyn Hutchinson's *The Ecological Theater and the Evolutionary Play*,[8] a book that inspired my grad school ponderings, and reflect on how today, loaded Winchester in hand, I'm truly an actor on that stage.

But what's with this heaviness? I intend to slay a mammal, as do rattle-snakes and other predators I've studied for decades. I've killed smaller animals for science. And although we're not exactly trapping prehistoric prong-horn in a canyon cul-de-sac, what could be more part of nature than taking responsibility for what we eat? Concerns that the deck is stacked against our quarry moderate during three hours of chilly immobility, as I settle into the winter landscape and contemplate the job at hand: superimposing the rifle's quarter-inch, V-shaped rear sight with its metal-dot front sight on the deer's six-inch kill zone, from almost a football field away. I can't flinch the trigger pull despite knowing that a lipstick-sized bomb will ignite next to my right eye, despite knowing that less-than-perfect performance would leave a wounded animal to suffer.

Zero tolerance for error? How about just being a tagalong, watching the others hunt? But then I'd be only a spectator, no longer a participant. . . . At this point the stand feels like an anechoic chamber I entered years ago while lecturing at the Bell Labs—heartbeats were audible in that "place where sound goes to die," as the host described it, and now I strain to hear my *lub-dup, lub-dup* in the frigid morning air.

A gunshot at 7:48 proves to have been Greg's scoped Marlin .30-30, a modern variant of my rifle, and when we meet up at 10 he teaches Tracy and me how to follow a blood trail several dozen winding yards to the dead buck. At first things aren't obvious, but Greg patiently urges us to cut back and forth over the rocky ground, attending to red flecks on pebbles and dry oak leaves. We soon catch on, and although I've seen countless live and road-killed whitetails, the one before us is riveting: Greg's bullet ripped right through the thorax and caused rapid fatal shock from blood loss. I ask how many he's killed and how many rounds were used. "Same number," he says without a trace of bragging, "fourteen or fifteen total." From this and other comments, we're getting the picture: These guys *never* take bad shots. They don't fire if branches would deflect a bullet, the angle of aim isn't favorable, and so forth. They eat every animal killed. Greg deftly eviscerates the deer in amazingly short order, and we lug it out to the nearest ranch road and David's pickup.

Longhorns graze among us back at the Chateau. Truth be told, I came as much for them as for the hunting. Domestic herbivores usually seem tame and needy, but these animals exude charm and independence. Descendants of stock released in Mexico more than five hundred years ago, shortly after the Columbian landfall, Texas longhorns are famous for their namesake weaponry and ability to thrive in hostile circumstances. The Double Helix cattle do seem well adapted to this rugged terrain, as if they belonged here; as nineteenth-century rancher Charles Goodnight exclaimed, "Evolution! If I could take longhorns and breed them into the best herd in America in eleven years, what could I do in eleven million?"[9] Moreover, longhorns are flat-out handsome, their colors "more varied than those of the rainbow," as folklorist J. Frank Dobie noted.[10] David is devoted to the breed, maintaining an informational website and writing articles for ranchers' magazines about the genetics of coat colors and horn conformation.[11]

Today, though, the oldest cow in David's herd has a wickedly suppurating infection of black, pink, and yellow over her left hip. He hadn't the heart to shoot her on his last visit, but now he waits until Brilliant Mary is looking away and fires his 7mm magnum into the base of her spine. The old gal crumples as if struck by instant narcolepsy. He wonders softly if her twitching is reflexive, shoots again, and she goes still. Next he fixes a strap around her neck, Greg and Cody hold her horns up, and they tow the cow behind the truck several hundred yards into the brush. Then David saws off the head so as to save her skull. We've watched two suffering mammals dispatched within twenty-four hours, plus Greg's buck, all killed respectfully and without fanfare. After eggs with venison chorizo, tortillas, and coffee, we make a fast trip to town for food pellets with which to bribe the cattle into a corral tomorrow.

As our first day at the ranch rolls on I'm increasingly attuned to the songs of insects and birds, hoping it'll warm up enough for Strecker's chorus frogs to begin calling. By 3:45 P.M. I'm in a ground blind constructed earlier this afternoon out of old boards among a few scruffy oaks, closer to Lower Pond than its namesake tree stand and within shooting range for me and the Winchester. A doe flushed into woodland above the pond's far shore while I was walking down here, and now I wonder if there are others in the vicinity. After an hour, a great blue heron alights and hunts the shallows— pausing, peering, moving—as if to underscore the afternoon's ironic tranquillity with a question or two. Can I be as effective as the natural-born killers who have long held my interest, let alone carry out the task with equal grace? I scan for deer and marvel over the lush, gray-green lichens that coat the ubiquitous oaks.

At 5:40 there's tan movement on the distant hillside that might be a bobcat, but through binoculars I see a doe bend to nibble, take a few steps, feed again, and walk slowly to the pond's dam, where she stops and looks toward me. Next she drops below the earthworks and moves a bit faster, parallel to and below them, disappearing in vegetation at the near end to my left. I raise the rifle and pull the hammer to full cock, then assume a comfortable rest position against a tree limb. My body feels tight all over, and I silently affirm this won't be my only chance; I must breathe correctly and never make a bad shot, even if that means not taking one. Within a minute the deer walks along the shoreline and stops about sixty yards away, head up. I sight just behind her shoulder, slowly let out half a breath, and squeeze. As the Winchester explodes she wheels sharply left and leaps awkwardly out of view. I've been taught to sit tight, allow the deer to lie down and lose consciousness instead of pushing her into running farther, but now I'm anxious about whether the shot was good. Darkness is falling rapidly, so I lever in another round, put the rifle on half-cock safety, and zigzag in her direction.

Forty yards from where she was hit, the doe lies on her right side, head up as if caught napping, and there's no response to my approach. White belly fur glows like snow in the flashlight beam as I recall dazed looks of the gravely injured people I treated as a young medic, then zoologist George Schaller's comment about a zebra, mortally wounded by African wild dogs and appearing more witness than victim of its fate.[12] I note a bloody half-inch exit wound behind the doe's left shoulder and ponder another shot, but her head sinks and she doesn't react to a tug on one hind leg. Back up the ridge I find David, who offers congratulations and help with field dressing. The deer's entrance wound turns out to be mid-abdomen, meaning either she turned away at the last instant or I flinched, and there'll be gut damage. "Not a problem, we'll clean her up," David says, handing me his knife. "She died fast, that's the key."

I pinch up and slice a patch of skin at the edge of the rib cage, then zip open her abdomen. After slitting skin forward to the neck, I push and pull hard through the chest skeleton, slice all around the diaphragm to free viscera from the body wall. Reaching in, I sever trachea, esophagus, and major blood vessels. Everything is wet and warm, slick and membranous. Even the pastel guts and brick-colored liver are shockingly vivid. Cutting around the tail end of the intestine turns out to be the trickiest part of the procedure, during which I recollect Greg sharpening his knife and wonder if self-inflicted stab wounds are common among hunters. Then we roll the innards out onto the ground, leave the doe belly-down to drain, and go for David's truck. Within half an hour she's hanging by hind legs from a tree outside Chateau Vert.

The five of us dine on pheasant from David's recent bird hunt, along with tenderloins from Greg's deer and a tasty New Mexican posole Tracy brought, all savored with several good cabernets. Before crawling into my sleeping bag I head out into the icy night to take a leak. Coyotes are chorusing off to the east, under stars and a sickle-shaped moon that would have made the Spanish poet García Lorca proud. I surmise they've discovered the new gut pile.

Approaching Lower Pond well before dawn, I notice the eyes of a deer shine yellow-green in nearby brush and disappear. By 7 A.M. a rusty-orange glow paints surrounding vegetation, highlighting a dainty pencil cholla cactus with red fruit among the knee-high prickly pears. Soon sunlight blasts over the far ridge into my eyes—not good for hunting, so I swivel and scan for wildlife. Scattered white blotches against khaki slopes emerge through binoculars as grazing longhorns, and I hear faraway lowing. Later, walking back to camp, I admire the upraised tail flags of four spooked deer, and at the Chateau a lifted hunk of sheet metal yields a handsome black-and-straw–striped Texas patch-nosed snake, which we photograph and release back under its lair. All in all, it's one of those mornings for smiling with no need to ask why.

After brunch we move the longhorns into a corral and segregate nine calves for their first visit to a veterinarian. Roundup consists of me driving the pickup over gravel two-tracks and honking while David sits on the tailgate rattling a sack of food pellets at converging cattle—more reminiscent of a rural Fourth of July parade than *Lonesome Dove,* but I'm enthralled.[13] Cody and Greg cajole the calves with gentle hand slaps and "Come on now," separating them through iron chutes and gates into a trailer. Safety entails not getting pinned among the adults, their pecking order arbitrated by the pokes and prods of massive horns, so one of us waves them back while another rations pellets into feeder buckets. In town, a jovial vet vaccinates the calves, David freeze-brands them with copper irons dipped in liquid nitrogen, and one calf gets castrated, all with surprisingly little commotion. Among the youngsters, I'm drawn to four-hundred-pound Cinco de Mayo, a black-headed brindle with butterscotch eye rings and eight-inch horns who's slated to become a Double Helix herd bull.

I occupy the Chateau stand in late afternoon, see nothing, and enjoy the blanketing darkness as I amble back to join the others. We dine on fresh deer tenderloins, garnished with Tracy's pickled shiitakes and washed down with a Chilean carmenere, and David answers my questions about ranching specialty cattle in a market that maximizes corporate profits on feedlot beef. Longhorns rarely need help birthing or lose calves to predators, unlike Herefords, he notes, and their meat is leaner, all consequences of a free-living existence. His Brilliant Mary produced more than twenty calves, a phenomenal output compared with that of "improved" European breeds and presumably thanks to the regional environment's role in shaping her genetics. Later I drift off to sleep imagining Cinco de Mayo at three-quarters of a ton, siring calves on a landscape far more biologically diverse than typical pasture.

Walking together in the dark our third morning, Greg and I acknowledge with silent nods that last night something, probably a coyote, dragged his deer's gut pile several yards. I hunt the Streamside ground stand, decked out in a camouflage ghillie mask and pestered by quarrelsome fox squirrels. Evidently I don't actually look like a lichen-encrusted stump and everyone in the ravine has my coordinates: nobody shows up. Later, however, Greg recounts how a buck followed a young doe all around him, then made foot-stamping threats from so close that he threw up his arms to spook the deer away.

Back at Lower Pond in the afternoon, I see no game but relish another encounter with the longhorns. Three cows with different patterns of black or brown splotching on white are off to my left, and three respectively colored calves converge on my right. Each momma's bawling is distinctive to my ear, as are their offspring's shorter but more plaintive responses, and after the predicted pairings, in which calves go to cows rather than vice versa, they all resume grazing. At first their munching sounds right next to my ears, then trails off as the herd drifts toward water. I'm reminded that a comparably heavy shrub-ox, said to have preferred hilly country, was among North America mammals lost to extinction at the end of the Pleistocene. After dinner and a spirited game of Liar's Dice, I fall asleep thinking about how conservationists typically bemoan cattle as a blight on the landscape, whereas this place feels *wilder* for the longhorns, enhanced rather than diminished by their presence.

Well before light the next morning I'm hiding in the Rock Pile, where Tracy got a buck yesterday and others have seen ringtails, striped skunks, and

What, indeed, is wild? Cinco de Mayo (shown here as a three-year-old, *top*) with Shonuff and their calf *(bottom)* are the legacy of five hundred years of natural selection in arid North American landscapes. On this Texas Hill Country ranch, longhorns coexist with more than a hundred species of native vertebrates. (Photos: H. W. Greene)

porcupines. She nailed a perfect chest shot and followed the blood trail before summoning Greg for help with field dressing. Now there's heavy frost on everything, and despite thick camo pants and thermal underwear, the rickety metal chair reminds my skinny butt of branding irons dipped in liquid nitrogen. Dawn breaks as I'm wondering whether Paleo-Indians worried about poorly thrown spears and suffering prey, immediately doubting the relevance of such things for hungry primates, be they chimps or people. We regroup at 10 A.M. and proceed to Upper Pond, where two pintails and a green-winged teal fall to shotguns and end up on the grill for lunch.

By 3:30 on this second shortest day of the year I've returned to Lower Pond with one of David's Rugers. Concentric ripples break the shiny water surface and provoke boyhood memories of fishing with a cane pole. Crickets trill softly. At 5 the sun's dazzling orb is low on my right and a tidelike demarcation creeps up the far slope, with deep shadows below and a late-afternoon glow on the foliage above. Chilly breezes caress oak twigs by my face as two common ravens flap and croak like haikus across the crystalline December sky.

Ten minutes before shooting light ends, a young antlered buck walks broadside without pause, followed by a big doe that gives me perfect vantage. I'd just been lining up the scope's crosshairs on cobbles near the pond, practicing holding steady, and as my trigger squeeze starts, the bucolic scene goes freeze-frame and grainy, like an antique glass-plate photo, then shatters in cascades of image shards and lingering echo. *Ka-POWWWWW,* and a couple of seconds later something crashes softly a few dozen yards down to the right. As before, there was no perceived recoil, although my rifle carries a punch on the target range, and the Ruger even more. I walk straight ahead for a hundred yards or so, turn in search of blood spatter, and see the fallen animal facing downhill toward me. As I reach the doe David arrives for more tutelage in field dressing, and when creamy fluid trickles from my artless nick of her udder, he notes that the fawns have long since been weaned.

Our last morning I'm still taking it all in. Near where the deer fell are crimson dribbles on grass stems, amoeboid splotches on chunks of quartz, and garish streaks on a prickly pear, as if someone ran upslope juggling a bowl of precious liquid. Dropping to my knees, I contemplate bloody Rorschachs on rock lichens that rival the tragic beauty of a Goya painting, and dismiss smug hopes the doe didn't suffer, convinced only that her five-second death was easier than one dealt by coyotes and likely kinder than the lot I'll one day draw. Walking back, I meet Greg, who's shot a feral pig at the Plunge Pool, and we discover six fetuses while field dressing it. David is concerned about

White-tailed deer killed by an omnivorous, bipedal primate with a fire-spear, Terrell County, Texas. (Photo: H. W. Greene).

these destructive invaders and glad that this first sow killed on his ranch won't give birth, but still we marvel at her wild boar–like canines and bristly pelage. I'm also intrigued by the inch-deep layer of gray-white fat, reminiscent of a body sock, which bolsters claims of snakebite not affecting pigs and thus their efficacy as enemies of venomous serpents.[14]

By noon David and I leave the others, who will also be departing to join their families for Christmas, and take my deer to town for processing.

Two nights later, in the Philly airport's aptly named Terminal F—where Ithaca-bound travelers often mutter, "We're screwed, flight's canceled"—I spot Brian Aquadro, a colleague's son on holiday leave from the marines. A fine writer and photographer who likes reptiles, he's carrying a Jim Harrison novel I'd recommended. Last summer he'd admired my old carbine when we spent a morning plinking; now, when I mention the does and tell him I'm looking to buy a scoped, flatter-shooting rifle, my young friend feigns indignation. "Why not practice offhand, without a rest," he advises with a smile, "get more confidence, so no matter the circumstances, you'll shoot well with that Winchester?" My main goals are to eat healthy meat and never make bad shots, but I also crave some sort of primal validation, and Brian sounds wise beyond his years.

Homeward bound at last, I conjure up the last few days: Strong coffee and hiking by flashlight; a dawn chorus of winter birds and chattering squirrels. Porcupine latrines at the Rock Pile, littered with turds that look like wooden cigars. Whitetails gliding across clearings as afternoon light dimmed, their heads dropping to browse and swooping back up, ears flared. In the plane's dark privacy I think about those marvelous sensibilities destroyed by my selfish, violent acts, as if a curtain has ripped open, revealing an empty stage. I remember raw flesh smells and steaming intestines, cracking ribs as the knife jerked along, then shattered lungs and hot sticky liquid, the anatomical feel of tracheal cartilage. Within hours we were raising our glasses to the deer, their grilled tenderloins perfect with a bottle of Los Vascos Reserve, and I'd begun reflecting on life as a born-again predator.

Over the next three years I read widely, annually refill our freezer with venison, and conclude that hunting is philosophically murky.[15] Much of what is said about its ethical pros and cons presumes we no longer *need* to hunt, for example, but because I'm an omnivore and repulsed by industrial meat production, *someone* has to kill the farm animals or wild game I eat. Sport hunters focus instead on notions of "fair chase," some of them disgusted by Texans "baiting" with corn, yet those critics include southerners who plant "deer crops," northerners who "drive" deer, and westerners who set up ambush near scarce water and "rattle" antlers, enticing their quarry with the clatter of sexual competition. Weaponry and age-affected skills further confound these matters—some of my arrow-wielding friends disdain rifles but occasionally lose wounded animals, and in fact, even were I capable of doing so, what could be more "fair" *and yet terrifying* for a mammal than to be run to exhaustion and slaughtered with a tomahawk?

Turns out things get even more complicated. As a child I unsuccessfully lobbied for keeping the runt piglet from a litter of Grandpa's Durocs, and was captivated by a roadside zoo's Hampshire shoat that performed tasks on cue. As a field biologist I've enjoyed many encounters with collared and white-lipped peccaries, the New World relatives of Old World true pigs, and wart-hogs are among my favorite African animals. In May of 2011, however, despite all that porcine karma, I'm back at the Double Helix to hunt descendants of Eurasian wild boar. On the flight to Austin this loomed as an exciting yet unsettling prospect, because although my new Ruger .270 shoots where it's pointed, the awkward truth is, I really like pigs.

As if a plague of hogs tearing up the countryside weren't enough, the preceding summer ushered in the worst drought in Texas history, and folks are hoping we're not in store for a string rivaling the brutally dry 1950s. In wetter years the Hill Country sparkles like a Cézanne painting, resplendent in blue-bonnet, paintbrush, and other flowering forbs, but now it's parched brown and dusty, devoid of bright colors save the occasional yellow-blossomed prickly pear and an *Echinocereus* cactus with violet petals. Only northern cricket frogs are calling, there's not a snake in sight, and last week David lost his first calf ever, perhaps due to weather, since there were no signs of predation. Three cows have healthy newborns, though, so along with hunting pigs, I'll revisit my ethological roots and watch longhorns. Cinco de Mayo has been moved elsewhere, but I soon recognize two other favorites and wonder how they're fitting in. McTavish is a classy white heifer with India-ink speckles, named after David's grad student who studies Longhorn genetics, while half-brother Placido, all red-brown splotches and dark eye rings, looks like a bovine jester.

The week unfolds and I indulge in natural history. Out here sights, sounds, and smells are explicit, everything is signal and nothing is noise; vistas appear framed in my mind's eye, and things happen that I've never seen before. A turkey vulture lands at scum-covered Lower Pond and drinks like a barnyard chicken—quick pecks, followed by gawky naked head tilted skyward to swallow. Later I'm sitting quietly, watching a browsing doe from 150 yards out, when, within seconds of wind gusting against my back, her head swings up, ears spread wide like parabolic reflectors, and she laser-locks my gaze for a long minute before sauntering off. Another morning McTavish and Placido are play-jousting as the herd grazes around Ann and David's new house, yet by sunset even the spindly-legged calves are at Upper Pond, having followed their mommas a couple of miles uphill despite temperatures in the nineties.

My own wanderings are freeform, conducive to daydreaming. Because Galápagos iguanas lack enemies and are deemed "island tame," are they no longer "wild"? Might hunting, viewed by some as a cruel perversion, enhance wilderness by promoting fear and alertness? In any case, feral pigs, a genetic hodgepodge of domestic and wild boar lineages, are center stage this trip. Their rooting is everywhere obvious, and David recently dropped two standing side by side with a single bullet. Tuesday a sounder of adults and piglets crosses the power line cut where I killed a buck last December, then a smaller group trots off as I approach Upper Pond, but in the afternoon rumbling thunder and grisly memories of a long-ago lightning victim convince me to take shelter. Wednesday a nearby storm drives us in early again, but by 10:30 A.M. skies have cleared and telltale diggings have replaced corn at the Lower Pond feeder, so off I go, in search of the culprits. After an hour traversing the incoming ravine, back at water's edge, I surprise a flock of wild turkeys, perhaps those that gobbled as I first walked down here in the predawn light.

By late Wednesday afternoon I've seen pigs on three occasions but never had a good shot, and I will leave Friday morning. As still another thunderstorm skirts the Double Helix, David goes for a stroll near the house, scares up a sounder, and returns carrying a dead black shoat. I'm impressed by the youngster's ragged five-inch exit wound, having puzzled over why external gunshot damage to deer looks minor compared to that on some humans I handled as a medic. We chat over beers and admire a glorious sunset while David guts, skins, and grills the little boar, concluding that maybe people and piglets are simply fragile compared to a whitetail's chest and hide. During the ensuing feast our conversation veers from tactics for teaching about biological diversity to a good-natured debate about wild pigs as ecological proxies for Pleistocene flat-headed peccaries.

Thursday morning I park the pickup a quarter mile from Upper Pond and hike in, discover the feeder is bare of corn, and surmise that pigs have been here too. David has encouraged me to hunt the drainage ravine, so I set out along its near slope with the .270 slung over my shoulder, striving for silent footfalls on the pebbly ground. This proves tricky with aging knees, each step a reminder of how much more quietly other large mammals move. Every few minutes I detour into the ravine, check out its nooks and crannies, then press on. I try to visualize how a pig will materialize in my consciousness, wonder whether any have already spooked. By 11:30 my stomach is grumbling, so I cross the dry creek bed and work back toward the far end of the pond, which I'll check out before heading back to the truck.

Oaks, mesquites, and junipers dominate the open terrain. At rock outcrops I watch for black-and-white banded crevice spiny lizards but see none, perhaps because it's overcast and too cool for the ten-inch-long sunbathers. I round a brush-choked boulder pile as the game trail leads up a grassy knoll, thinking it looks good for broad-banded copperheads and puzzling over why we haven't run into any western diamondbacks. *Perhaps it's just too dry, or maybe hogs really are hard on snakes?* These herpetological daydreams have barely registered when sixty yards ahead a black pig traverses the hill, evidently having come through fencing off to the right. It pauses, a curved dark mass with head to the ground. There being no discernible wind, I creep forward and gain cover behind a gnarly old oak, from which I intend to kill my first feral hog.

Just as the Ruger reaches shoulder level, my attention jerks fifty yards hard right to a dingy brown sow, plus a black shoat like the one we ate and three gray piglets, all bobbing toward me in grass on the other side of the fence. *What the hell, I'm surrounded by pigs!* When I turn back there's a more formidable-looking, reddish-gray animal, with prominent woolly ears and dull cream facial markings, a dozen yards to the left of the black pig and facing it. This hog immediately turns toward me, advances a couple of steps, and growls loudly—*These things growl?*—at which point the black pig looks at its compatriot. My vantage is from downslope, and although I've encountered no cattle this morning, I cannot be sure a longhorn isn't behind the pigs, in my line of fire. Truth to tell, I am also worried the largest hog will charge, because it's still staring my way and growling. During these few seconds I steady the scope's crosshairs between its eyes, intending to fire when the pig takes just one more step forward. Instead it wheels out of sight, the black one rockets off in a blur, and I hear muffled snorts, then silence. The five that were approaching from behind the fence have vanished without a sound.

Hoping to catch up, I sneak into a side ravine and up onto the pond's dam, but see only a soft-shelled turtle basking on the far bank. I sit under a tree for fifteen minutes and scrutinize the surrounding slopes with binoculars: nothing appears, and it's lunchtime. Back at the house David barely conceals incredulity, given his resolve to knock back hog numbers. "Shoot first, then ask questions," he says with a grin, scrambling eggs with chorizo while I throw together black bean quesadillas, intent on heading back out and hunting until dark. It's almost midnight when, explaining my hesitation with the red hog, I tell of how as a nine-year-old I fired Grandpa's .22 at a stray cat and inadvertently killed one of his chickens. Then I announce that, finally, late this afternoon, I reduced the Double Helix pig population by one.

The sow's erect, rounded ears and coarse black pelage first put me in mind of a short-legged bear instead of the desultory, sparsely haired pigs of my youth. For five hours I'd hidden under some oaks, scanning a meadow near Upper Pond. Time slipped by while cardinals whistled, little dark butterflies fluttered among cactus flowers, and a trio of young raccoons traipsed through the grass. Late-afternoon hues shifted from golden brown to soft bluish-gray. She'd appeared at the clearing's edge as a mourning dove *oWoo-woo-woo-woo*ed, and from 125 yards away this manifestly self-willed creature looked like fresh charcoal, as if her fur could soak up the waning daylight. After several minutes the sow walked two dozen yards and stopped behind a small juniper that blocked my shot, but through branches I glimpsed her head tilting this way and that. *What is this big gal thinking?* A few more minutes passed before she moved into the open again and paused, oblivious, as I centered low on her shoulder and reminded myself not to flinch.

Then an orange blast cleaved the dusky silence, and having practiced shooting with both eyes open, I saw her collapse, kick the sky, and go still.

Evening had cooled nicely when we suspended the field-dressed sow from a tree near the house. David guessed her weight at 150 pounds, about a third of which Kelly and I later converted to delicious ribs and pulled pork. What I hadn't counted on, never mind an invasive species humanely harvested for food, was the wave of sadness that tempered my elation, a vivid reminder that we, alone among predators as far as we know, empathize with prey. This led to reading in Juliet Clutton-Brock's *A Natural History of Domesticated Mammals* that wild pigs are omnivores—hunter-gatherers like us, not browsers or grazers—and their young require more prolonged parenting than those of deer and cattle.[16] Then I remembered the bears Gordon Burghardt was studying when I first moved to Tennessee, how impressed he was by their broad diets, maternal care, and other similarities to humans. Conflicting emotions aren't surprising, I concluded, given the fuzzy-eared sow's ursine countenance and my lifelong affection for her kin.

As if fondness for pigs and some heavy-hitting hunting literature hadn't challenged my evolving notions of "wild" and "natural," returning to Texas a couple of months later further upped the ante. Because a piddly five inches of rain had fallen on the Double Helix over the past ten months, all the drainages and Lower Pond were dry; Upper Pond was shrinking fast, its surface a variegated sheen of red and green algae. The deep-rooted mesquites owned their usual emerald glow, whereas many oaks were brown or had already dropped leaves. Driving in I'd seen two turkey vultures tugging on a cardboard-thin raccoon mummy, but the arid-adapted longhorns seemed unaffected, forsaking baled hay to graze elsewhere on the ranch. One afternoon they showed up around the house and gorged on purple cactus fruits, as if the huge bull Quannah or his matriarchal counterpart, Buffalo Springs, had suddenly declared, "Okay, herd, let's go out for prickly pear!"

Although I'm comfortable killing game for food, butchering seemed dubious with the porch thermometer showing 103 in the shade. Pigs, however, compete with native species for scarce resources, so at David's request I would humanely dispatch them whenever possible. Strolling about the first three days I encountered as many as twenty at a time and tried to imagine how much water a ton of hogs might drink, but they always bolted when I was still hundreds of yards away. Once a noisy dove spooked the shoat I was about to shoot; another time, creeping behind trees, I spotted a couple of adults and assorted smaller pigs beside Upper Pond, but they caught my smell at 250 yards and trotted off. Late on the fourth afternoon, though, I snuck up on their encampment below the dam. A mottled orange shoat ran squealing past my feet before rejoining the melee, and as half a dozen animals streamed off down the ravine I dropped the last one, a young sow. Then I sat quietly for a few minutes, admiring her long wild-boar head, her horizontal tusks, and her sooty gray, mud-caked fur.

Twelve hours later some sixty black and turkey vultures occupied a snag over the sow, which lay untouched except for a few flies and carrion beetles. After three days, only a pig-sized stain and hundreds of prickly pear seeds from her stomach marked the spot, the nearby skeleton and scraps of fur testifying to my role as carcass-providing carnivore. Then things took a bizarre turn with discovery of two shoats mired at pond's edge, with sludge over their eyebrows yet ears up, as if alive. A vermillion flycatcher sallied overhead, lending color to the morbid scene. As I moved closer the shoreline liquefied underfoot, implying the pigs' fate was a mud bath gone bad; they weren't bloated or smelly, nor were there signs of struggle. That evening, because swine wallow, David and I discussed whether the shoats could have somehow been holding their breath. Next morning, my first glimpse through binoculars revealed a dark cloud flapping free of the muck and squabbling over bloody chunks of pork. Vultures were eating the drowned pigs from above, like macabre picnickers spooning the flesh out of watermelons.

Almost four years since that first trip to the Double Helix and all this ragged complexity must be admitted, especially that my experiences are tame compared to stalking elk with a longbow, let alone spearing a woolly mammoth. I've shot eleven deer and two feral pigs, not Faulkner's mythic bruin in *Go Down, Moses*,[17] and when everything's said and done, there's blood on all our faces. Bragging rights are in the eyes of the beholder—the domain of aesthetics rather than ethics, Cornell philosopher Jim Tantillo points out[18]—and as my Texas sportsman-conservationist friend Robert McCurdy says, beyond healthy meat and clean kills, what hunters most stand to gain is an honest, personally illuminating experience. While I'm on provocative topics, though, let's inquire into mouth-watering barbeque from aquifer-polluting pork factories, of prime rib *au jus* from urine-soaked feedlots; let's keep in mind the biological richness of well-managed ranchlands compared to soybean fields. And let's continue asking what it means to be part of nature, redouble efforts to keep the ecological theater well staffed, and promote a richly acted evolutionary play.

Meanwhile, I'll keep practicing offhand with the Winchester but hunt with my newer rifle, under circumstances that make filling our freezer with environmentally healthy, humanely harvested meat more likely than not. And I'll never forget walking up on that first doe at Lower Pond, when instead of cosmic revelations my mind simply flashed *sacred*—an improbable draw from an agnostic's vocabulary—followed by *intimate* and, finally, *grateful*.

FOURTEEN

Field Biology as Art

THESE PAGES OPENED WITH BAYARD Taylor's allusion to a princess attacked by robbers for the jewels she wore, his metaphor for Sierra Nevada streams ravaged by nineteenth-century gold miners. Let me draw to a close with a better-known, twentieth-century quote. As Senegalese environmentalist Baba Dioum said, "We will conserve only what we love, we will love only what we understand, and we will understand only what we are taught"[1]— words that nicely epitomize how research plays into both education and conservation. Field biologists observe organisms, discern patterns, and determine their causes; then, ideally, society uses that knowledge to coexist with those with whom we share the planet. And as I've tried to show, in the course of extricating nature's secrets and passing them on, we gain personally as well.

Introducing herpetology students to desert reptiles was a welcome assignment, but my most rewarding teaching at Berkeley entailed the class Joseph Grinnell initiated decades earlier, in which Henry Fitch had assisted as a grad student. Like Bob Stebbins, the previous Museum of Vertebrate Zoology herpetologist, I taught natural history of the vertebrates with an ornithologist and a mammalogist. My lectures summarized amphibian and reptile diversity, then illustrated anatomy and ecology with local examples. I explained how frogs and salamanders projected their tongues, acted out my "arms-and-torso" model for how a snake eats. Most importantly, every spring we got several dozen undergraduates paying attention to their surroundings, and they in turn shaped my life as an educator.

During weekly laboratory sessions we laid out specimens for students to distinguish some hundred species of vertebrates. They learned turtle shells and bird feathers, memorized rodent teeth and other minutiae. In one lab, with approval from the campus animal welfare committee, I fed a mouse to a rosy boa, so they could see the snake's jaws pulling over its meal as well as confront the reality that predators kill and prey die. Beforehand I mentioned the rodent was surplus from a research colony, described what would happen, and emphasized that students weren't required to watch. Then I gently maneuvered the mouse within range. The serpent spun its muscular coil, and after a few seconds both were still. We discussed snake jaw mechanics while the meal was swallowed. In my twenty years of teaching that course, no one ever left the room or complained.

Every week our class carpooled to an East Bay regional park, each trip having a specific goal, such as observing the breeding behavior of California newts or the foraging of shorebirds. We were out until noon, in small groups supervised by TAs and professors; the undergrads wrote Grinnellian-style field notes.[2] We visited some sites repeatedly, so they could contemplate seasonal changes. Students also conducted field projects and summarized them in scientific publication format, on topics ranging from mundane to bizarre—studies of brown towhees and gray squirrels on campus were routine, but one young man carried road-killed cats and dogs into a meadow, sat for hours under a cardboard blind, and watched turkey vultures find the carrion.

Our kids came inappropriately decked out for the inaugural field trip in bright clothing and fancy shoes, and as the morning wore on they fumbled with borrowed binoculars, grumbled about rain, and were bewildered at the prospect of distinguishing dozens of little drab birds. By the last trip, however—one to a ranch in a nearby arid valley—they'd transformed into seasoned naturalists, confidently sorting rodent skulls out of barn owl pellets and calling classmates over to see a western rattlesnake among some rocks. That course, in particular the field trips, has no doubt influenced thousands of people since its early-twentieth-century beginnings. A rural gas station attendant once grinned through Stebbins's windshield and blurted out that thanks to Bob he'd never again ignored wildlife. At graduation, parents exclaimed that "natural history changed my kid's life"; I'm still getting glowing feedback long after leaving Berkeley.

For decades Stebbins had given the first and last lectures of the course, but with his retirement ornithologist Ned Johnson assumed the lead. Henceforth Ned would cover the introduction, and we'd share the final session. My closing remarks, initially titled "Many Happy Days I've Squandered," were illustrated with photographs of animals in faraway places.[3] One year, though, a student came to my office and said he wanted to find out "what it all means."

Surprised by his candor and gravity, I mumbled something about seeking one's own path, shifted to a lighter topic—and resolved to develop a better answer.

From then on I would end the class with slides from our field trips and comments about what makes life worthwhile. First are relationships with friends, lovers, family, and so on. Students always chuckled when I admitted no special wisdom on that front, saying that I'm fumbling along in the agony and bliss like everyone else. Second is being creative and feeling accomplished, a sort of fulfillment that nature can provide. Finding a rattler in its ambush coil next to a rat nest is precious, as is the pleasure of identifying "little brown job" birds and feeling food items through the belly of a live snake. Lastly, we need time, space, and grandeur to shrink our problems, humble our perspectives. "Think about it," I'd say. "Across centuries of strife and for millions of years before our hairy ancestors strode onto the African savannas, salamanders have been surfacing when it rains and hiding out when it doesn't, indifferent to our much-vaunted importance. Surely there is a sense in which those amphibians, among the most abundant of vertebrates, transcend us."

Then I would remind my listeners that judging from aboriginal rock art, Pablo Neruda's poetry, and countless other examples, being inspired by nature is an ancient, universal aspect of humanity.[4] At this point I'd be looking out at dozens of familiar faces, stars of insightful and humorous incidents on field trips. My voice would crack with emotion, while a few students sniffled audibly. We'd stood around a muddy pond and sexed Pacific chorus frogs by checking for the dark throat of males or eggs visible through the abdominal skin of females; we'd walked a ragged line across meadows, an arm's length apart, searching for rattlers, then gasped when a golden eagle swooped so close overhead that binoculars were irrelevant. Now, in the interest of lightening things up, I'd throw out an anecdote. My friend Ben Dial had an overhead transparency, I told them, peppered with hundreds of random black dots and boldly titled "The Universe." He'd flash that sheet up to his biology class, point to an off-center speck labeled "You are here," then pause for effect and announce with a big grin, "That's us!"

I would close the lecture by emphasizing that nature is disappearing fast and needs clean air and water, as well as protection from excessive impact—just those ecological circumstances that Grinnell devoted his life to studying and that we had hammered away at all semester. I'd note that for the final exam students should remember how snakes crawl and why woodpeckers have long tongues; they should understand fertilization in amphibians and lactation in mammals. But more than that, I hoped that years later they'd recall that those thousands of species of frogs are different in ways that matter, and that they'd still pick apart owl pellets and identify food items. I would sign off urging them to share these pleasures with their friends and children, and to keep nature in mind when they vote. Since moving to Cornell I've concluded introductory biology and herpetology classes with versions of that talk, and its underlying message informs all my teaching.

If ever a group of organisms exemplified Baba Dioum's dictum linking research, education, and conservation, it's dangerous snakes. They encapsulate problems of living with animals that might kill us, as well as our general reluctance to care about slithering creatures—after all, empathy *is* a stretch when it comes to animals without fur or feathers, the more so when they lack limbs and eyelids. The good news, however, is that turkey vultures and mountain lions are easy if people can appreciate rattlesnakes, so I'm optimistic. Although rattlers have declined in many areas and some are threatened with extinction, we've learned a lot about their ecology and behavior, and that knowledge plays an ever larger role in our coexistence. Early on in the natural history course, I'd noticed that facts help students appreciate deadly serpents, but I wanted more: I wanted people to care about snakes as they care about birds. And I had an idea.

In 1991, Pulitzer Prize–winner Natalie Angier visited my Berkeley lab for a *New York Times* piece on pitvipers. First I explained how radio transmitters revolutionized snake biology, permitting herpetologists to study the predatory and social behavior of these secretive animals. As we looked over my captive rattlesnakes, maintained for teaching, I pointed out that many of us have abandoned the old macho way of handling them, that manually restraining a snake's head resembles a predatory attack and causes it to struggle violently, risking injury to both animal and researcher. Instead, we use a shepherd's crook–like "snake hook" to gently prod the serpent into a plastic tube, such that it can be carefully grasped at midbody, with the front end safely inside the cylinder. One of the payoffs of humanely observing rattlers is that they are revealed as fascinating animals, exemplifying far more than just their namesake defensive adaptation.

While Natalie gingerly touched a tubed Great Basin rattlesnake's buzzing tail and marveled at the velvety feel of its gray and charcoal-brown skin, I told her my pipe dream: that ecotourists would visit timber rattlesnake dens.

Natalie was open to the beauty of snakes and keen on public education—"Pit viper's life: bizarre, gallant, and venomous" soon ran in the Science section of the *Times*[5]—but the idea of *seeking them out* must have sounded preposterous, because even she responded with irony: "Ah, yes, get my travel agent." I was only sorry that because timber rattlers are restricted to the eastern United States I had no prospects for setting up such a trip.

Western rattlesnakes still frequent the Berkeley campus, but I chose timbers for their prominent cultural and conservation status. John Smith mentioned them in his 1621 *Map of Virginia, with a Description of the Countrey,* and Revolutionary War flags with iconic rattlers warned "Don't tread on me." In a 1775 *Pennsylvania Journal* essay, Benjamin Franklin opined, "Her eye exceeds in brilliance that of any other animal and she has no eyelids. She may therefore be esteemed as an emblem of vigilance. She never begins an attack, or, when once engaged, ever surrenders. She therefore is an emblem of magnanimity and true courage. She never wounds until she has generously given notice even to her enemy, and cautioned against the danger of treading on her. The poison of her teeth is the necessary means of digesting her food, and at the same time is the certain destruction of her enemies. The rattles are just thirteen—exactly the number of colonies united in America. One of these rattles, singly, is incapable of producing sound; but the ringing of thirteen is sufficient to alarm the boldest man living. She is beautiful in youth and her beauty increases with age; her tongue is also blue, and forked as lightning, and her abode is among the impenetrable rocks." [6]

Although surely many of Franklin's three million fellow colonists disliked timber rattlers, they couldn't cause irreversible declines in a widespread, abundant species—yet scarcely two centuries later, with our population a hundred times larger, those elegant animals are endangered or extirpated in parts of their range. This predicament stems from a collision between snake biology and human prejudice, although timbers rarely bite us and are important natural predators. The females are adapted to low mortality and slow population turnover: in the Northeast they require a decade to mature, breed two to four times in a twenty-five-year lifetime, and bear about ten large young per litter. These snakes are thus demographically fragile, and because they visibly aggregate at dens, marauding people have wiped out entire populations; in some regions only isolated colonies persist where rattlers were once common. Moreover, because they travel hundreds of yards between winter dens and summer hunting ranges, they need large chunks of continuous habitat—all in all, a textbook case of vulnerability.[7]

Fast-forward to June 2001, after I'd moved to Cornell and joined a regional conservation group. Finger Lakes Land Trust was purchasing Steege Hill because it encompasses a timber rattlesnake den, across the Chemung River from a population on the Nature Conservancy's Frenchman's Bluff Preserve. Now the Trust wanted to spotlight timbers in their Talks and Treks lecture series. I was finally getting my chance to promote rattlesnake ecotourism!

At a Thursday night gathering, I started by recounting the biology of black-tailed rattlesnakes in Arizona, where Dave Hardy and I had watched individuals for more than a decade, capturing many aspects of their behavior in photographs. I explained that the first image was of a rattler in threat display, ready to strike, but rather than acting *aggressive,* it was *re*acting out of fear at the close approach of a person. As researchers, I explained, we strived not to disturb snakes and therefore rarely saw defensive postures; thus the other photos would illustrate their complex, idiosyncratic lifestyles. Blacktails, like timber rattlesnakes, hunt mammals, lie around after big meals, search for mates and wrestle with rivals, court and mate, give birth, and attend their young. They visit certain places repeatedly within well-circumscribed home ranges. They occasionally and inexplicably climb trees. And, like the big male that bent a fern out of his future strike path, mentioned in chapter 1, they often surprise us.

After the slides, with the audience standing safely back, I hooked a timber rattler out of its container and onto the floor. Keeping an eye on the four-foot reptile, I explained that these snakes are easily avoided, accidents are rare, and with proper medical treatment even serious bites are survivable. Then I pointed out that there are more than three thousand species of snakes, some with lifestyles we can scarcely imagine. About 10 percent are vipers, with hypodermic needle–like fangs for injecting a cocktail of immobilizing, digestive venom; thus armed, vipers eat items exceeding their own weight, so imagine me consuming a 200-pound hamburger without benefit of hands or silverware! Two-thirds of vipers, including copperheads and cottonmouths, have infrared-imaging pits behind the nostrils, and of those pitvipers, thirty-seven New World species possess a noisemaker on their tails. The rattle is a marvelously interlocking set of hornlike segments, vibrated about sixty times per second by specialized shaker muscles and used to warn away enemies.

While our audience admired the demonstration rattlesnake's calm demeanor I described the species' natural history, based on pioneering studies by Bill Brown, Marty Martin, and Howard Reinert, as well as my postdoctoral associate Rulon Clark.[8] People responded to the live rattler with comments like "beautiful," "awesome," and "isn't it wonderful to be so close!" Their questions ranged from "Can you tell his age by the number of rattles?" to "Are rattlers evil?" First I tackled the easy one, noting that a segment is added roughly twice a year when a snake sheds its skin, but because old segments are worn off, counts don't correspond to age. Next I said that having been an ambulance driver and army medic, without special knowledge of theology, I regard evil as exclusively human. In the course of appreciating rattlesnakes, I speculated, we might contemplate violence and mortality without anthropocentric baggage, maybe gain a little clarity in such matters.

On Saturday morning a dozen newly primed snake enthusiasts joined me at Tanglewood Nature Center in Elmira, just north of the Pennsylvania border and adjacent to Frenchman's Bluff. Our leaders were retired ophthalmologist Art Smith and his daughter, Polly Smith-Blackwell, lifelong area residents who serve as preserve stewards. Over the next three hours they guided us to several rock outcrops under openings in the forest canopy. By early summer, males and nonbreeding female rattlesnakes have moved off into the woods to ambush mice and chipmunks, but gravid females bask close to their winter den, the better to maintain elevated temperatures for developing embryos. The first few clearings we visited yielded common gartersnakes, resplendent in black and yellow stripes, but no rattlers. As we approached the last outcrop, though, a couple of dozen yards out, Art's upraised hand signaled us to halt. Other than the leaders, no one in the group had ever seen a wild timber.

Art pointed toward the rocks. One by one our guests distinguished scaly loops from leaves and fallen branches. Soon these newly minted snake hunters looked like amateur ornithologists, except their binoculars were angled *downward* and they were whispering about a timber rattlesnake. Birds fluttered and sang overhead, but the people stared straight ahead at the snake. She was perhaps three and a half feet long, coiled in dappled sunlight on a jumbled stack of slabs; she had recently molted, judging from the lustrous yellow and brown crossbands. Her hindparts and tail were velvety jet black, and her abdomen was so swollen with pregnancy that stretched bluish skin was visible among the scales. We kept a respectful distance, and soon everyone stopped talking—just stood watching those immobile coils. No obstreperous TV personality dangled her by the tail and crowed about how she was so "dangerous." No one poked her with a stick. And that beautiful animal never so much as *chick-chick-chick*ed her rattles, as these snakes sometimes do when agitated.

The first Talks and Treks program was consistent with my experience teaching natural history at Berkeley: with preparation, people readily treasure live rattlesnakes. But we've got a long way to go. Fifty percent of all pitviper species might be endangered, yet only a few dozen have protected status.[9] Meanwhile, people slaughter rattlesnakes by the thousands at "roundups" in Oklahoma and Texas, and Missouri legislators tried unsuccessfully to exempt snakes from protection as wildlife. Until recently, terciopelos were deemed hazardous and killed around buildings at La Selva, despite only one serious snakebite there in more than forty years. And when I asked a Texas park ranger about snakes, she responded, "They're *bad* this year!" Think about that: she said they were *bad* . . .

Conserving venomous serpents inevitably depends on controlling their negative impact on us and vice versa, as is true for more popular organisms like elephants and big cats. Nevertheless, Dioum's testimonial about saving what we love and loving what we understand emphasizes how research and education are linchpins for appreciating less widely liked creatures. We can indeed all be teachers, in classrooms or over backyard fences. So if you agree with me that the Earth is wilder with dangerous animals, in ways that repay tolerance, tell your friends something good about rattlesnakes. And remember, information that helps us understand, love, and conserve nature follows from someone carefully attending to her particulars—someone like my high school mentor, the father of snake ecology.

Henry Fitch seemed physically fit during our longest interview for this book, only months after he'd fallen off an embankment into a chilly creek at the age of eighty-nine. Daughter Alice was up from Oklahoma visiting her parents at the K.U. Natural History Reservation, and the four of us talked all afternoon and into the evening. Virginia served a fine dinner of fried chicken, mashed potatoes and gravy, corn on the cob, and homemade yeast rolls. Almost eighty, she still had raven-black hair and a lively sparkle while recounting decades of Fitch family life. Topics ranged from my first visit with them as a sixteen-year-old to current projects. At a mention of "youthful indiscretions," I said, "Huh?" and Virginia explained that she'd married young and divorced the other man when he went overseas. "Then," she added, smiling and hugging Henry from behind his chair, "I met this wonderful guy!"

From time to time I checked my notes, and, in spite of his enthusiasm for this book, Henry was reticent. "Do you believe in God?" resulted in a pause, then "I have no religious beliefs. Although I was raised in that environment, natural history does it for me." I wanted to find out why he keeps turning over boards and catching snakes, what being *in* nature meant to him. At one point I blurted out something about finding peace as problems shrink in the face of grandeur and diversity. "Sounds good to me" was his only response, with a soft chuckle and perhaps a hint of irony. I was also curious how Henry decided what to record, because since the 1930s he'd gathered data for which initially there were no conceptual frameworks. In 1949 he'd outlined in *Ecology,* then as now a highly respected journal, *what* to write down, but almost nothing as to why specific details were valuable.[10] Optimal foraging theory had its debut only in 1966,[11] inspiring behavioral ecologists to consider parameters that Henry began measuring thirty years earlier with no theory whatsoever.

So I kept coming at him from different angles, hoping Virginia and Alice would jump in with something definitive or nudge Henry for answers. Asked about favorite habitats, he attributed a preference for deserts to the M.V.Z. Nevada trips, when he found exciting reptiles and enjoyed the open views. Queried about favorite animals, he responded, "Alligator lizards, copperheads, and gartersnakes, because of their interesting natural histories." By the end of the evening the best responses I could get for the deeper questions were "my initial interest in zoology was innate" and "I began to write down everything that interested me."

Two years later, back on the Reservation, Henry was audibly winded as we crested a limestone ridge, explaining without self-pity how he'd lost stamina but hoped to complete one more field season. Otherwise the man seemed no different physically than on my last visit, and at ninety-one his hair was light brown. He walked stooped and wore brown, thin-soled work boots, khaki pants, a faded blue jacket over a Berkeley herpetology t-shirt, and a black baseball cap. Cotton snake bags, custom made by Virginia, hung through his belt, in case we needed to bring anything back for note taking. He used a walking stick with a nailhead protruding on the bottom to steady himself, turn over cover items, and probe matted grass for pieces of tin he'd laid out to attract snakes. As our conversation turned to current projects, Henry spoke with fatherly pride of a paper he'd written with Alice, about changes in tree species diversity over the past fifty years.[12] Once prairie, the Reservation had lost a third of its fauna due to fire prevention, lack of grazing, and forest encroachment. "On the bright side," he added, "there are still bobcats as well as the occasional timber rattlesnake, even a black bear recently." Back at the

Henry Fitch in 1991, trim and fit at age eighty-one, examining a common gartersnake, Fitch Natural History Reservation, Douglas County, Kansas. (Photo: V. Snider)

house he showed me a good-sized cedar, planted many years ago as a tiny family Christmas tree.

On the K.U. campus that night, Henry sat in the front row for my lecture titled "Heroes, Theories, and Organisms as the Central Focus of Biology." I held aloft his fifty-year study, *A Kansas Snake Community,* and summarized the man's phenomenal legacy of animals marked and measured, stomach and scat contents tallied, and papers published.[13] Henry had bridged the descriptive natural history of Wallace and Darwin, I explained, in what amounted to three careers, counting the San Joaquin Experimental Range, K.U. Natural History Reservation, and tropical expeditions. In fact, although best known as a herpetologist, he'd made contributions to mammalogy exceeding those of many nominal mammalogists. During my talk, to exemplify the importance of field observations for generating new research opportunities, I described island cottonmouths feeding on fish regurgitated by seabirds,[14] and afterward he asked about their litter size. Then, with head tilted down and a slight swing of the chin, he added with a chuckle, "Oh, and thanks for the plug!"

A prairie moon glowed through low, heavy fog as I drove back to town after taking Henry home; orange lightning flashed over fields and hedgerows while I wondered anew, *What makes him tick?* Years ago, when I'd complained about scarce funding, he replied, "I've always spent time on whatever interested me most—with or without grants—and have greatly enjoyed all my projects, especially the fieldwork." More recently he told the author of a book on Kansas personalities, "I wouldn't change a thing. People who work with animals in the field, whether snakes or birds or rodents or monkeys, find it deeply satisfying and wouldn't trade it for any other career—even though it may not be financially rewarding."[15] And Henry once wrote Alice that "if, as a young person, I'd dreamed about my future and the world I'd like to see, it would have been about the same as the life I've had. Getting a Ph.D., having a loving, supportive wife, children like you and John and Chester, grandchildren like Tyson, Lena, and Ben, living on the Reservation, teaching natural history and studying reptiles, including anoles and pitvipers, and making two dozen trips to nine countries in the tropics have all been great experiences."

My hero and friend passed away at Alice and Tony's home, a few months shy of his hundredth birthday. Just weeks earlier, when her father asked about watersnakes he might catch in nearby creeks, Alice replied, "Sure, Dad, but then what?" Henry's response—"We'll mark and recapture them"— implicitly answered some of my questions as well. His inborn curiosity and work ethic flourished in graduate school, and later, having landed a job that played to his strengths, he needed no further justification. That gift for field biology and stubborn self-confidence must have been obvious to Grinnell in 1931, when a new student arrived in Berkeley, fresh off an Oregon pear ranch. Henry's long, happy life was inseparable from his quest to understand nature.

Three gifts of pleasure and revelation, in the form of contemporary literature, bolted out of the proverbial blue and into the home stretch of this book. The insights they conveyed, along with the reasons I so strongly disagree with one author, help frame my portrayal of field biology as art.

I'd devoured Larry McMurtry's Pulitzer Prize–winning *Lonesome Dove* and admired the television adaptation, in which Robert Duvall stands out as Gus McCrae, among his many fine performances. Both versions played to my affection for westerns with far grittier nostalgia than the Zane Grey yarns and John Wayne movies on which I was raised, and they touched on troublesome themes to boot. A cowboy's death from cottonmouth bites, improbable in the details, captured snakebite's horror; and never mind Comanches and desperados—my favorite character was more troubled by his relationships with women and the changes wrought by his own invasive culture. I vowed to revisit the novel someday but instead fell into watching the six-hour miniseries every few years, until Caroline Fraser ended a *New York Review of Books* essay about McMurtry with a quote from Gus I immediately coveted.[16] In two sentences, that old rascal had encapsulated my favorite works by Albert Camus, Jim Harrison, Mary Oliver, and Gary Snyder.[17] But I couldn't use his words without seeing their context, and surprisingly—because I often mark such things—a thorough thumbing of my copy revealed no dog-eared pages.

Course now clear, I dived back into McMurtry's masterpiece and, perhaps for having reached Gus's age, relished it even more the second time around. I was also uneasy. Colorful characters, a couple of cute pigs that eat rattlers, some startlingly violent incidents, and six hundred–odd pages later, what if I couldn't find that quote? Then one night, well into the cattle drive that occupies much of the story, I noticed a folded page corner. The two-inch-thick book had been so tightly packed for my move to New York that the dog-ear was flattened into obscurity, but now its bent page tip beckoned powerfully. I strained to savor the moment, not rush to what I hoped lay just ahead.

Gus and a teenage cowboy were riding apart from their herd and suddenly found the ground strewn with bleached animal skeletons. As I turned that page, Newt, orphaned as a child, was lamenting a good man's violent passing, asking about life's meaning—and there was the quote. "The earth is mostly just a boneyard," Gus drawled, perhaps chewing on a grass stem and daydreaming of his beloved Clara, "but pretty in sunlight."[18]

The Abstract Wild, by philosopher and mountaineering guide Jack Turner, requires a bit more explanation.[19] In the summer of 2005 I'd brought his slim volume along to reread at Fort Clark, Texas, while working on this book within a vulture's soaring gaze from where *Lonesome Dove* begins. Seven years earlier, Turner's manifesto had been so irritating that I'd tossed it across my office in disgust. I'd been *uneasily* angry and couldn't now remember why, so I spent the better part of a day dwelling on what the author himself calls a rant, then finally went for a stroll along Las Moras Creek to clear my head. Two tropical green kingfishers zoomed by, reminders that although the Chihuahuan Desert begins only a few dozen miles to the west, my 1880s stone dwelling sits near where the Edwards Plateau's southwestern edge meets a mesquite savanna that extends up from Mexico. I spotted a diamond-backed watersnake basking on an overhanging tree limb; dropping low to minimize my profile in the late afternoon sun, I inched closer for a photograph before the four-foot-long serpent slid into the current.

After an hour or so of creekside idyll, I had to concede that defensiveness clouded my earlier assessment of Turner. I am an academic, after all, and having labored hard in the service of conservation as well as spent lots of time outdoors, I resented his sweeping condemnation of my profession. I also knew that if I were more single-minded I could do more; if I were stronger, more agile, and had more stamina, I could go farther. And I knew that the loss he laments, the disappearance of something bigger than us that we cannot control, is tragic. So, setting aside professional and personal insecurities, what are the issues at hand?

The Abstract Wild is a carefully argued tirade, 125 pages of white-hot rhetoric in the service of ideals that must resonate with anyone who's spent time in the backcountry. Turner begins with a surreal discovery of ancient rock art in a remote Utah canyon, one whose entry required harrowing, highly skilled maneuvers, and he decries that because now the site is more readily approachable, its aesthetic values are diminished. I have only to recall my initial resentment at registering for a hiking permit and packing out feces in Paria Canyon to empathize with his point. Farther into the book Turner eloquently recounts the impact of large carnivores on our psyches, then berates biologists who study bears, lions, and wolves for depriving those creatures of their wildness. He wants as much land as possible to be inaccessible to all but the few humans capable of entering it with negligible impact. Leave nature alone, he asserts, and she'll take care of herself. Treat predators with respect, and they won't eat us. Wildlife management, he believes, is the spiritual enemy of wilderness.

This is powerfully engaging stuff, but Turner is too self-righteous, too quick to stereotype biologists as soulless hacks and too facile in dismissing science. What *The Abstract Wild* lacks is anything more than an occasional dash of humility, any admission that others deserve their own routes to wildness, any recognition that education and research influence conservation. It would have been only minimally fair to admit that without science there'd be no California condors and black-footed ferrets on Earth today. Instead Turner admiringly quotes John Muir (though not his preposterous claim of never having glimpsed a drop of blood in nature);[20] he touts as models of peaceful coexistence writer Doug Peacock's field time with grizzlies and the Kalahari San people's relationships with lions[21]—all of which I admire—but fails to admit anything like jaguars that prey on forest Indians and reticulated pythons that attack Philippine hunter-gatherers,[22] or prairie rattlers that bite hapless children and kill beloved hunting dogs. Most folks simply won't tolerate dangerous creatures over the long haul unless we effectively face up to their deadly potential.

Despite my misgivings, a wonderful irony emerged as I absorbed Turner's account of a pathetic Asian zoo where, after watching visitors toss seeds through cage bars at a mountain lion, he flies into a rage and is physically restrained from throttling a man teasing the cat. Ben Dial would have done the same, I thought to myself—and in the next moment I realized that Ben would have *loved* this book. He would have brushed aside its arrogance, *admired* Turner's unabashed dogmatism, and forced me to acknowledge the admirable values underlying what I still see as narrow-minded and smugly critical. I turned a few more pages and found a faded yellow charge slip, squinted at the tiny blue print, and was brought up short by the date: I'd bought *The Abstract Wild* the night before I flew to southern California and reached my old friend hours before he died. Ever since then, with a nod to Ben's memory, I have recommended Turner's brilliantly provocative tome to students.

Rock art also played a role in the third gift, a book by anthropologist Carolyn Boyd that rescued me from the awkward task of justifying my subtitle. I'd started thinking about art and science as I left Texas for a Ph.D. at Tennessee, when Bill Pyburn handed me an essay by Gunter Stent.[23] The Berkeley molecular biologist claimed that those two endeavors are fundamentally similar, pointing out that although anyone might have untangled DNA, elucidation of its structure in a particular manner reflected the specific quirks and talents of James Watson and Francis Crick. This was an act of collective *creativity,* and something in the whole enterprise was *theirs.*[24] At about the same time I began to recognize that as a field biologist I was discovering and conveying facts about nature as well as clarifying my own life. Observing, describing, and interpreting—the practice of natural history— amounts to expressing realities as I perceive them, in publicly and privately rewarding ways. That in turn sounds a lot like what painters, musicians, and poets are doing.

Scientists to whom I've mentioned *Tracks and Shadows*'s subtitle have said they liked the sound of it, whereas artist friends were skeptical about linking field biology with their craft. I'd arrived at my summer writing retreat correspondingly insecure, wondering if I could penetrate a dense literature on the meaning of art and mollify their concerns. Soon after I met my landlord, though, he urged me to head out to Seminole Canyon State Park and Historic Site, seventy-five miles west of Fort Clark, and learn some archeology. I made a mental note to do just that, stopped in a bookstore, and within a few hours was delighted to find that Carolyn Boyd's *Rock Art of the Lower Pecos* had eased my way.[25] Her task was to interpret paintings at five sites clustered around the confluence of the Pecos River and Rio Grande, left there by Archaic hunter-gatherers more than three thousand years earlier; in so doing she ran head-on into the ambiguity of "art." I figured that seeing some of those paintings firsthand, with Boyd's analysis fresh in mind, would help me better understand the problem.

Fate Bell Shelter is upstream from Boyd's sites in Seminole Canyon, and on a sweltering July morning I followed a park ranger into the gorge, thence midway up a dark limestone bluff and under the high rounded ceiling. We looked out on scummy-green pools and scrubby oaks in the canyon bottom, wondered out loud about hunting with atlatl and spear on the surrounding, thinly vegetated Edwards Plateau. Then he noted the shelter's murals, dominated by life-size maroon anthropomorphs (humanlike figures) with exuberant head ornaments. He explained that they'd been painted with mineral pigments plus a binding agent from deer bone marrow and an emulsifier from yucca root. Among dozens of other designs on the curving rock surfaces, although some are faded or otherwise difficult to identify, a few have defining characteristics. Dark red color half fills a vertical feline and streams from its mouth, as if the cat were losing blood or perhaps its soul. Near one end of the shallow cave, successively more tilted, S-like markings terminate in an obvious snake, from whose jaws emanate lines I suppose signify hissing.

Fate Bell's motifs affected me in the manner of fossils, as windows on the past, and I couldn't resist visualizing their ancient milieu. The climate was moister and the landscape dominated by grasslands instead of desert, so perhaps pronghorn were common, as they are today a hundred miles to the west. Rock art, however, exerts an emotional allure beyond the primeval, because we know someone very much like us made it. Soon the shimmering clear skies and absence of modern sounds enhanced my sense that the shelter had been recently vacated, that instead of several thousand years only a few days separated the ranger and me from its earlier occupants. Occasional chirps from grasshoppers penetrated our contemplation, and in the intervening stillness I imagined Archaic people whispering around us, scampering off into the canyon. Under such circumstances, who couldn't empathize with

Natural history, now and way back then: *(top)* Fate Bell Shelter, an Archaic site in Terrell County, Texas, and *(bottom)* the image of a bleeding cat from the shelter's back wall. (Photos: H. W. Greene)

the creative stretch implied by those anthropomorphs and that upended bleeding cat, or feel some timeless unity in our human search for meaning?

Nonetheless, as Boyd explains, much of what's been done with paints, charcoal, and other media, including prehistoric pictographs as well as some of the world's most famous paintings, doesn't qualify as art under certain present-day definitions. "Fine arts" arose as a French phrase in the mid-1700s, and the English word "art" was first defined with a narrow, contemporary meaning in 1880; art as abstractions to be viewed, evaluated, and appreciated is an even more recent, almost exclusively Western formulation, coincident with disdain for utility. As Boyd points out, this leads to a motivational discord between aboriginal artists, who produce works for function and power, and collectors who buy and display them for strictly aesthetic purposes. Her analyses suggest the Pecos River panels served adaptive roles for small-scale foragers by facilitating communication and, via drugs derived from peyote cactus and other plants, as portals to their spiritual world. If these ancient representations aren't art, she concludes, we need a more flexible and inclusive concept of that term.

Even as Fate Bell Shelter shored up my confidence in the notion of field biology as art, I recalled some of the world's oldest paintings, discovered in 1994 in a French cavern and made more than thirty thousand years ago. They include astonishingly accurate renderings of animals, and in one of them a Pleistocene cave lion, with evident scrotum and lacking a mane, walks crouched beside a smaller female—typical courtship behavior, as carnivore expert Craig Packer noted upon viewing the Grotte Chauvet murals. For whatever reason, the ancient cave painters accurately recreated what they saw for others to contemplate, and in so doing left us the earliest recorded examples of nature study, rivaling those of "the best naturalists of our own era."[26] Boyd's Pecos rock art commentary and Packer's analysis of the Chauvet images convinced me that science and art have sometimes literally been one and the same.

It's time to admit that my final vertebrate natural history lecture didn't fully answer the student's question about life's meaning, to step farther out on that philosophical limb. Many biologists find no necessary conflict between science and religion, although some are convinced the former can answer everything, while others are born-again zealots. I'm inclined toward the first group, out of parentally ingrained tolerance and because there remains the vexing matter of how *did* it all start? On that last count, as my writer friend Chuck Bowden says, science still comes up a few bricks shy of a full load, and I've simply walked away from the question, plodded on through Ugandan rainforests, Mexican barrancas, and Brazilian cerrados. Like Ben Dial, Henry Fitch, and Ed Abbey's mountain lion, amid doubts, failures, and catastrophes, I'll never bawl over my soul. Besides, those wily Texas alligator lizards are waiting in Hill Country ravines, and I've got some new snake projects down that way too.

One might still ask, though, isn't there more? In an indifferent cosmos, are there grounds for hope? We are in an increasingly tight spot, our quality of life, even our existence, threatened by climate change, pollution, extinctions, wars, starvation, and pestilence. The state of the Earth looks ever more dismal; there are way too many of us consuming way too much, and unless we effect major changes the consequences will be dire for many species, including us. Besides reducing greenhouse gases, connecting habitat patches, and better managing our food, water, and energy consumption, we desperately need to get people into the wild—even if, for some of them, only in ways that are timid by Jack Turner's standards. In a world of seven billion people, many wretchedly poor and starving, hitching our hopes for nature to exclusionary ideals is dangerously counterproductive.

We must encourage engagement with nature, especially among urban youth, through activities like photography, bird watching, fishing, hunting, and wildland gardening—a term coined by tropical biologist Dan Janzen for conservation in human-dominated landscapes, especially when practiced by locals.[27] As far as revealing previously unknown details, Fitch's work confirms that expensive, high-tech gear isn't essential; nor are journeys to distant lands prerequisite, as shown by my late colleague Tom Eisner's elegant research on insects in his Ithaca backyard.[28] Anyone with a notebook and willingness to pay attention can practice valuable natural history, and field biologists should help others sharpen those basic skills. We also need to make observations part of a universally accessible record. Digital cameras, GPS units, and other modern accoutrements as well as web-based archives will help, but we can begin without them.

Bad news is all around, so I leaven it with success stories. *American Naturalist,* thanks to my former student Jonathan Losos's stint as editor, now publishes natural history along with the conceptually focused papers that have long been its stock-in-trade. Turns out the ivory-billed woodpecker might not be extinct,[29] and recently a Korean schoolteacher discovered a spectacular new salamander, its closest relatives in Italy and California.[30] Venomous snakes are still widely maligned, but thanks to Natalie Angier the *New York Times* ran another engaging, well-informed piece on them.[31] Fitch studied Kansas timber rattlesnakes into his nineties, finding one he'd marked twenty-four years earlier,[32] and people in my part of New York are ever more actively conserving those velvety pitvipers. Likewise, Ontario has education programs about massasaugas, and Arizona firefighters perform science-based rattler and Gila monster relocations, all in the interest of protecting dangerous reptiles. And amazingly, grassroot protests in Georgia, Oklahoma, and Texas against rattlesnake roundups might yet crush these grisly spectacles.

For those seeking hope on continental scales, rewilding won't be easy, but it might be doable. After all, wolves are once again taking elk out West, and recently the largest coyote I've ever seen walked through my suburban New York backyard—indeed, recent research shows our visitor carries some wolf genes picked up during her ancestors' trek eastward.[33] As I look to the future, with luck and lots of hard work, bolson tortoises are coming back in the northern Chihuahuan Desert,[34] and maybe someday elephants also will return to their Pleistocene New World haunts. I smile to imagine folks traveling from all over to show their kids a wild hundred-pound turtle, and wonder if in a few dozen generations big gray pachyderms with flapping ears and long trunks, exposed to the selective environs of a North American steppe, will look more like mammoths than they do now.

In craving hope on a local scale, I often recall that my friend Ben claimed his best years came after his heart transplant. He *didn't* surrender to hardship and he *did* smell the flowers, so I strive to honor his example, savoring simple pleasures and small triumphs. One night the owner of the West Texas cottage in which I first drafted this chapter took me to the Owl's Nest Café, a little roadside joint just east of the Pecos and north of the Rio Grande. As beers were handed all around, a huge fellow in a weathered straw hat introduced himself as "Pooch" and revealed that he cowboyed on a nearby spread. "Uh, no last name?" from me brought only leaden silence, then, "Nope, just Pooch." When he mentioned the ranch, though, I remembered a gate sign near Baker's Crossing, on the Devil's River, and my reserve gave way to curiosity about local natural history. Queried about skinny reptiles with small legs, Pooch said, "Yep, I've seen some scaly-looking, long-tailed things with kinda pointy heads, creepin' under leaves and roots." That sure sounded like Texas alligator lizards, and this guy worked right where, in 1854, army surgeon Caleb Kennerly caught the "large lizzard in bushes" that a Smithsonian zoologist used as the type specimen of this species.[35]

Pooch didn't seem inclined to speak beyond his knowledge, so with the northernmost distribution of a favorite tropical serpent in mind, I asked if he'd seen any big black snakes on the ranch. The cowboy finished a long swig, and I ducked as his massive arms swept wide for emphasis. "We catch sight of a *fourteen-footer* every month or so, real shiny like, seems to be workin' the banks of the creek," he exclaimed, "and I'm here to tell you ... whut's yore name? Oh, yeah, Harry . . . man, that's a big feller!" Indigo snakes reach eight feet in Texas, I thought to myself, and ten in Guatemala, but fourteen feet strained credulity.[36] Nonetheless, at Baker's Crossing I'd seen Rio Grande leopard frogs and blotched watersnakes like those an indigo would hunt, and I had sightings from elsewhere in the Devil's River drainage. Under the circumstances there was no doubt about the monstrous serpent's identification, and I wasn't inclined to challenge this guy over anything having to do with size. Besides, his response to my next question had us grinning and clinking our longnecks in a toast.

"Course we didn't *kill* that big black dude, hell, why *would* we?" Pooch scoffed. "Our boss won't even let us shoot them ol' diamond-backed rattlers anymore, except when they're right around the foreman's house, 'cause he's got kids and all."

"Here's to fourteen-foot indigos," I replied, raising my bottle, "and cowboys who let those ol' diamondbacks crawl on by." Yes indeed, simple pleasures and small triumphs.

I am so fortunate to have a job studying nature, and with more than half my years spent, here's an answer for those who ask what it all means. I believe that like all living things, we are for a while and then we aren't, that Camus was right: life is how we find it and what we make of it. Each of us leaves behind tracks and shadows, but inevitably they fade and we end up one with the water and the rocks. Rather than push that view on others, though, I urge students to engage the wild in whatever form they find it, contemplating other realities against which to measure themselves. My own travels have entailed terror, pain, and death, but I've also relished the fragrance of drenched rainforest soil, been dazzled by silhouettes on desert canyon walls. I've marveled at big cats and rattlesnakes up close and personal, welcomed the sheltering solace of friends and loved ones; I've watched curious children peer into a bird's nest and beam with innocent delight. And along the way I've discovered beauty, mystery, and always, sooner or later, hope.

May of 2012 has me back on the Hillises' Double Helix Ranch, and what a difference rain makes! Gray corpses of three-hundred-year-old oaks dot ridgetops, testimony to last year's devastating drought, but healthy trees still populate slopes and valleys. Where everything was dead brown, now swaths of yellow rock coreopsis flowers outshine lime-green groundcover, their brilliant monotones punctuated by crimson firewheels, lavender skeleton-plants, and a few bluebonnets left over from April's peak. The tiniest streambeds are burbling, and both ponds are full. We walk miles and spot none of the wild pigs that earlier wreaked such havoc, suppose they've scattered with water more widely available, and enjoy watching Cinco de Mayo, the new herd bull, tending his cows and calves. Several whitetails bound off, plump and tasty looking, and the vermillion flycatchers and black-chinned hummingbirds are perky as ever. Best of all, given David's worries that many herps have perished from lack of moisture, we find five species of frogs and a half-dozen lizards, some bulging with eggs, as well as a sleek six-foot coachwhip.

Making a living inspired the Chauvet and Pecos painters' attention to their surroundings, but just as surely, like all natural historians, they noticed far more than was necessary for subsistence. I'll wager, too, that their attentiveness started as youthful curiosity, and one morning when I'm out on the Double Helix, searching for snakes and toting my Winchester in case any pigs show up, a rambling daydream gets me smiling. *Hope is a double-edged notion—freighted with life's uncertainties, we yearn for friendly outcomes—and what could be more hopeful than coming back to the Texas Hill Country, where this seven-year-old met his first bluebonnets and longhorns, first western diamondback?* Later that same day, as a pastel dusk softens the countryside and frogs and katydids commence their familiar nightly songs, it strikes me that nature is more like a bandit queen than a princess, worthy of love and protection but scornful of idolatry. Let's cherish our earthly boneyard, so pretty in sunlight, with no white flags!

NOTES

CHAPTER ONE

1. Taylor 1891, 155.

2. In the early 1980s I drafted from memory an account of my childhood through discharge from the army, then revised it based on feedback from friends and relatives, as well as from news articles. I changed a few people's names and minor details to protect their privacy. My field journals (see Herman 1986; Canfield 2011) archived in the M.V.Z. were invaluable for reconstructing events from 1978 onward.

3. Abbey 1988, xii.

4. See Greene 1988a, 271, for a photo of peccary hooves and sloth claws in jaguar scat.

5. See Greene et al. 2002 for blacktail field studies, including parental care.

6. See Greene 2003 for the chipmunk trail incident.

CHAPTER TWO

1. Wallace 1869.

2. Conant 1958.

3. Species are denoted by italicized binomials ("two names"), in which the first word, capitalized, refers to the larger category to which it pertains, its genus, and the second, not capitalized, refers to its species—thus *infernalis* is one among several species of the genus *Gerrhonotus*.

4. Smith 1946.

5. See Greene 2005 and Fleischner 2011 for discussion of and citation to other literature on the value of natural history.

6. My account of Henry relies on his publications (especially Fitch and Echelle 2000), as well as conversations and correspondence over more than half a century; correspondence with and about him by Grinnell, Hall, and others filed in M.V.Z.;

conversations with Virginia Fitch, Alice Fitch Echelle, and Tony Echelle; an edited volume celebrating Henry's eightieth birthday (Seigel 1984); and posthumous tributes by Duellman (2009), Henderson (2009), Hillis (2009), Pisani (2009), Plummer (2009), and Seigel (2009).

7. Van Denburgh 1922; Camp 1916.

8. See Stein 2001 for an outstanding biography of Annie Alexander, including the early M.V.Z. history.

9. See Miller 1964, Herman 1986, and Stein 2001 for biographical information on Grinnell; and Sunderland 2012 for an overview of collections-based research at M.V.Z.

10. See Sunderland 2013 for a history and overview of that course.

11. See V. Fitch 1984 for Henry's publications through 1983.

12. Fitch 1935.

13. Ruthven 1908, 3. His gartersnake opus frontispiece, labeled *Thamnophis ordinoides elegans* from Tule Lake, Oregon, is actually a gopher snake (*Pituophis catenifer*). I've found no mention of this elsewhere, nor have several Ph.D. graduates of the University of Michigan I've queried known of it. Henry recalled asking Jean Linsdale about the error, but the older man refused to speak of it.

14. Van Denburgh and Slevin 1918.

15. Fitch 1940.

16. Hubbs 1941.

17. Myers 1941.

18. Mayr 1941, 1942. Henry enjoyed showing him gartersnakes when Mayr visited K.U.

19. Fitch 1948a.

20. Rossman et al. 1996.

21. Stebbins 1954, 1966.

CHAPTER THREE

1. Wynn 1944.

2. Baker 1945; Ditmars 1931.

3. Oliver 1955; Conant 1958.

4. Greene 1961—first and only time this has ever been cited! The grotto salamander is now *Eurycea spelea*.

5. Anderson's (1965) *Reptiles of Missouri* was published posthumously.

6. Greene and Wakeman 1962.

7. Fitch and Greene 1965; Greene 1969a.

8. Cole 1966.

9. Fitch 1960a.

10. Klauber 1956.

11. Greene and Oliver 1965.

12. Blair's (1950) biotic provinces of Texas paper taught me that species have geographic distributions, with suites of species found throughout a region.

13. See Hajdu 2001 for the flavor of that scene, including Hester's role.

14. Irving 1981.

CHAPTER FOUR

1. Fitch 1948b; see Greene and Jaksic 1983 and Jaksic et al. 1981 for overviews of Fitch's predator studies.

2. Smith 1946, 57.

3. Fitch 1948c.

4. Fitch 1951.

5. Fitch 1963.

6. Fitch 1960b, 1961.

7. Fitch 1958.

8. *Anolis fitchi,* Williams and Duellman 1984; *A. duellmani,* Fitch and Henderson 1973.

9. Clark 1970; Plummer 1977.

10. Fitch 1960a.

11. Smith 1964, 460.

12. Savage 2002, 47.

13. Fitch and Fitch 1955.

14. Fitch 1999, v.

15. Hillis 2009.

16. Henderson 2009.

17. Savage 2002, 47.

18. Jaksic and Greene 1984.

19. Poran et al. 1987.

20. Seigel 1984, 213.

21. Fitch 1989.

22. Fitch 1999, 1.

23. Fitch et al. 2004; Fitch et al. 2003.

CHAPTER FIVE

1. Fitch and Greene 1965; Greene 1969a; Greene and Dial 1966; Greene and Oliver 1965.

2. Greene 1969b.

3. Generally quotes are to the best of my recollections unless sourced, but I instantly memorized that drill sergeant's comment.

4. That project (Greene 1973) provided my first inkling that anecdotes, such as that of the sand boa and jackal (Minton 1966), could help answer more general questions.

5. This quote from Wilson (1999, xi) speaks volumes about the discoverer of coralsnake mimicry: "At the dawn of field biology, Alfred Russel Wallace departed for the most distant and dangerous biotic frontiers of the world, carrying with him little formal education but a blessed love of reading and reflective solitude. He sought the insect-ridden Edens of which naturalist explorers dream. His principal lifeline to the English homeland consisted of specimens outbound—birds skinned, insects pinned, plants pressed—and sporadic payments for his treasures inbound. An intense young man, totally focused, awesomely persistent and resourceful, resilient to tropical diseases that killed so many others, and nobly selfless, even to Darwin, who otherwise might have become a bitter rival, Wallace endured, and he triumphed."

6. Harrison 1978, 197.

CHAPTER SIX

1. My account of Pyburn is based on conversations and correspondence during our thirty-year friendship, as well as information provided by his wife, Wanda; daughter, Karen Anne; and our friends and associates, especially John Karges.

2. Pyburn 1955, 1961.

3. Pyburn 1963.

4. Greene 1969c.

5. Greene and Dial 1966.

6. Eibl-Eibesfeldt 1970.

7. Tinbergen 1963.

8. Gehlbach 1972.

9. Schmidt 1932.

10. Later I learned that Shine (1977) independently used museum specimens to study snake diets.

11. Gehlbach et al. 1971.

12. Greene 1984.

13. Greene 1976.

14. Wallace 1867.

15. Wickler 1968.

16. Greene and Pyburn 1973.

17. Greene and McDiarmid 2005.

18. Darwin 1859.

19. Schaller 1972.

20. Klopfer 1962; Ewer 1968.

21. My portrayal derives from conversations and correspondence with Gordon over the past thirty years, as well as with his colleagues, students, and family. Among many publications, Burghardt's (1967) "Chemical Cue Preferences of Inexperienced Snakes" exemplifies the early snake work, while his 1992 study on human-bear bonding poignantly reflected on the bear studies. Among major contributions to ethology is a book on play (Burghardt 2005).

22. Greene 1994, 1999a.

23. Atz 1970.

24. Greene and Burghardt 1978; Greene 1979.

25. Hutchinson 1965; MacArthur 1972; Janzen 1977.

26. Kuhn 1962; Popper 1963; Dunn 1954; Taylor 1952.

27. Wilson 1975; Klopfer 1973, 1999.

28. Nisbett 1976.

29. Rand 1968.

30. Burghardt et al. 1978; Greene et al. 1978.

31. Herzog 2010.

32. Halberstam 1972.

33. "A world that can be explained even with bad reasons is a familiar world. But, on the other hand, in a universe divested of illusions and lights, man feels an alien, a stranger. His exile is without remedy since he is deprived of the memory of a lost home or the hope of a promised land" (Camus 1955, 5).

CHAPTER SEVEN

1. Davies et al. 2011.

2. Rodríguez-Robles et al. 1999. At the time of our study most nightsnakes were regarded as one widely distributed species. Based on Mulcahy 2008, those in much of California and Arizona are still called desert nightsnakes (*Hypsiglena chlorophaea,* including the animal figured on p. 87), but the one I found in Baja California was a coast nightsnake (*H. ochrorhyncha*).

3. Rodríguez-Robles et al. (2003) summarized Stebbins career.

4. Eakin 1975.

5. Stebbins 1966.

6. Earth Works Group 1989.

7. Fitch 1940.

8. Luke 1986.

9. Carothers 1986.

10. Greene and Jaksic 1983.

11. Jaksic et al. 1981.

12. Holmberg 1957.

13. García Márquez 2003.

14. Dixon and Wright 1975. These lizards are now placed in *Microlophus* and sometimes referred to as Pacific iguanas.

15. Greene 1982.

16. Losos and Greene 1988; Greene 1986.

17. Gould and Lewontin 1979.

18. Gould 1977.

19. Wilson 1975.

20. Mayr 1983.

21. Tierny 1987.

22. Flannery 2002.

23. Gould and Vrba 1982.

24. See Larson and Losos 1996 for a review of these issues.

25. Losos and Greene 1988.

26. French 1977.

27. Wright and Wright 1957, 260.

28. See Greene 1988b, fig. 8, for an image of that coralsnake and its defensive display.

CHAPTER EIGHT

1. Ditmars 1939.

2. Janzen (1983) provides a wonderful introduction to Costa Rican natural history, as does Savage (2002), with emphasis on its herpetofauna.

3. Hundreds of species of small brown frogs once placed in *Eleutherodactylus* and known to most herpetologists as "some eleuth" are now placed in several, not always closely related, genera (Hedges et al. 2008).

4. Greene 1988a.

5. Greene 1989.

6. Rand and Greene 1982.

7. Campbell and Lamar 2004.

8. Phelps 2010.

9. Pitman 1974, 1.

10. Greene 1994.

11. Isbell 2009; Headland and Greene 2011; and citations in both for primate and snake evolution.

12. See Drees 2003 for a discussion of evil in nature.

13. Pope 1955.

14. Goris 2011.

15. Neruda 1965.

CHAPTER NINE

1. Skaroff 1971.

2. Burghardt et al. 1972.

3. Bonnet et al. 2002.

4. See Headland and Greene 2011 for references.

5. Isbell 2009.

6. See Isbell 2009 and Headland and Greene 2011 for snake and primate evolution.

7. Kellert and Wilson 1993.

8. Robbins 1980, 182.

9. Owen's quote from Bellairs 1969, 108.

10. Gans 1986.

11. Gans 1986; Lillywhite and Henderson 1993.

12. Regal (1966) labeled this behavior regional heterothermy.

13. Greene 1983a, 1988b, 1997.

14. Schwenk 2008 and references therein.

15. Greene 1983a.

16. Cundall and Greene 2000.

17. For a video of me explaining to ten-year-olds how a snake's head works, and them showing me a better way, see https://www.youtube.com/watch?v=Mm9h6KE-ZOk, accessed January 26, 2013.

18. Barker et al. 2012; Murphy and Henderson 1997; De Lang 2010.

19. Murphy and Henderson 1997.

20. Twigger 1999.

21. Shine et al. 1998.

22. Auliya 2006.

23. Fredriksson 2005.

24. Head et al. 2009.

25. Haverschmidt 1970.

26. https://www.youtube.com/watch?v=x8gAdEyVhao, accessed January 13, 2013.

27. Heymann 1987.

28. Rivas et al. 2007.

29. See Headland and Greene 2011 for references to primate-python interactions, and Sunquist 1982 for fatal deer encounter.

30. See Headland and Greene 2011 for references to primate-boa interactions, and Montgomery and Rand 1978, Greene 1983b, and Trail 1987 for boa natural history.

31. Rivas 1998.

32. De Lang 2010.

33. Headland and Greene 2011.

34. Hill and Hurtado 1996, 162.

35. Njau and Blumenschine 2011.

36. Mark Knopfler's "I Dug Up a Diamond," from *All the Road Running,* with Emmylou Harris.

CHAPTER TEN

1. See Headland and Greene 2011 for primate and snake evolution.

2. Larrick et al. 1978. Likewise, among Paraguayan Aché, "Most individuals survive snake bites after a few days of intense pain. Occasionally, the victim will lose a limb from the toxin, which destroys the flesh near the bite, and the wound may

become severely infected. Most adult males have been bitten at least once and after they survive a bite they are called paje, which roughly translates as 'magic.' The most common circumstance is to be bit after stepping on a snake while looking up into the canopy in search of arboreal game" (Hill and Hurtado 1996, 153).

3. Kramer 1977, 752.

4. Colwell 1985.

5. Picado 1931, 14.

6. Campbell and Lamar 2004.

7. Campbell and Lamar 2004.

8. See Reilly et al. 2007 for foraging strategies in reptiles.

9. Cundall and Greene 2000.

10. Greene 1988b.

11. Savitzky 1980.

12. Cundall and Greene 2000.

13. See Pyron and Burbrink 2012 on the diversification of advanced snakes.

14. There is an enormous literature on venoms and snake biology. For earlier, still-useful overviews, see Savitzky 1980, Greene 1997, and Cundall and Greene 2000; for an entry into exciting new findings, see Mackessy 2010, Vonk et al. 2011, and Casewell et al. 2013.

15. For the functional and evolutionary morphology of fangs, see Deufel and Cundall 2006; Cundall 2009; and Vonk et al. 2008.

16. For an introduction to many of the world's vipers, see Campbell and Lamar 2004 and Phelps 2010.

17. See Klauber 1956 for the classic account of rattlesnakes; Greene 1992, 1997 for overviews of viper evolution; and Campbell and Lamar 2004 for coverage of New World pitvipers.

18. For elapid diversity, see Shine 1991; Spawls and Branch 1995; and Campbell and Lamar 2004.

19. Westhoff et al. 2010; Young et al. 2011.

20. Zamma 2011.

21. Shine 1991 is still the most accessible coverage of Australian elapids.

22. Spawls and Branch 1995.

23. Tiger keelbacks thus have both *venomous* and *poisonous* chemical weaponry; see Hutchinson et al. 2007.

24. Armentano and Schaer 2011.

25. Greene and McDiarmid 2005.

26. Rage and Bailon 2011.

27. Reinert and Cundall 1982.

28. Greene et al. 2002.

29. Klauber 1956, 737.

30. Anderson 1942, 215.

31. Wharton 1966, 154–155.

32. www.npr.org/player/v2/mediaPlayer.html?action=1&t=1&islist=false&id=120833365&m=120833151, accessed January 13, 2013.

33. Amarello et al. 2011.

34. *Tachymenis peruviana,* Greene and Jaksic 1992.

CHAPTER ELEVEN

1. Wright and Wright 1949, 1957; Conant 1958.

2. Gehlbach 1981.

3. See Greene et al. 2009 for a summary of this lizard's natural history.

4. Greene and Dial 1966.

5. Gould 1983; Dial and Fitzpatrick 1983.

6. Dial 1987a.

7. Dial 1987b.

8. See Jones and Lovich 2009 for summaries of the natural history of banded geckos.

9. Dial 1972.

10. Dial and Grismer 1992.

11. Wright and Wright 1957, 343.

12. Dial 1965.

13. See Merker and Merker 2005 for an overview of gray-banded kingsnakes.

14. Campbell et al. 1988.

CHAPTER TWELVE

1. Goin et al. 1978, 11.

2. Campbell and Lamar 2004, 268.

3. Greene 1992.

4. De Kay 1842, iii.

5. Skutch 1980, 256–257.

6. Skutch 1998.

7. G. Schaller, pers. comm.

8. See Lee 1996 and Nabhan 2003 for herpetological examples.

9. Marešová et al. 2009.

10. Burghardt and Herzog 1980.

11. Kiester 1997.

12. Matthiessen 2008.

13. Conant 1958.

14. See Ross 1989 for an excellent if outdated overview of crocodilians.

15. Pramuk 2008.

16. Brochu et al. 2010.

17. Ross 1989.

18. Kiester 1997; Greene 2005.

19. Burghardt and Greene 1988.

20. Andreadis and Burghardt 1993.
21. Burghardt 1997.
22. Griffin 1976.
23. Rivas and Burghardt 2001.
24. de Waal 2013; for video of the capuchin study and more, see www.ted.com /talks/lang/en/frans_de_waal_do_animals_have_morals.html.
25. Beaupre and Greene 2012.
26. Greene 1988b.
27. K. Wiley and J. Harrison, Kentucky Reptile Zoo, pers. comm.
28. Attenborough 2008.
29. Martin 2005 laid out his worldview.
30. Martin 1969.
31. Soulé 1980.
32. Janzen and Martin 1982.
33. Martin and Burney 1999.
34. Kurtén and Anderson 1980.
35. Donlan et al. 2005.
36. Donlan et al. 2006.
37. Rubenstein et al. 2006, 233, 235.
38. Rubenstein et al. 2006, 235; Packer et al. 2005, 927.
39. www.worldlifeexpectancy.com/country-health-profile/Tanzania, accessed January 17, 2013. Fewer than forty Tanzanians per year were killed by lions between 1990 and 2005 (Packer et al. 2005).
40. Rubenstein et al. 2007.
41. Van Devender and Bradley 1994.
42. Donlan and Greene 2010.
43. Bock and Bock 2000.
44. Truett and Phillips 2009.
45. Fitch 1999; Fitch, pers. comm.
46. Williams 2004.
47. See Callicott 2011 and Sandler 2012 for philosophical explorations; Manning et al. 2006 for "stretch goals" and "backcasting" in restoration ecology.
48. For journalistic coverage of Pleistocene rewilding, see Barlow 2000; Stolzenburg 2008; Fraser 2009; and Levy 2012.
49. Pringle et al. 1984.
50. Hardy and Zamudio 2006.
51. Moir 2006; Finkelstein et al. 2012.
52. There is a large literature on the controversial meaning of nature and how that debate affects conservation strategy—see, e.g., Soulé and Lease 1995; Janzen 1998; Nelson and Callicott 2008; Barnosky 2009; Marris 2011; Kareiva and Marvier 2012; and Noss 2012.
53. See Kay and Simmons 2002, Noss 2012, and references in both sources for divergent views on the ecosystem roles of Native Americans.

CHAPTER THIRTEEN

1. Ditmars 1931; Baker 1945.
2. Jeffers 1988.
3. "I never saw one drop of blood from all this wilderness" (Mighetto 1986, xxiii).
4. Beebe's (1927) bandit incident isn't corroborated by other sources (Gould 2004; Gould, pers. comm.), nor can I confirm my impression that he used a Winchester .30-30—photographs of Beebe with firearms usually show shotguns, but one includes among four weapons a lever-action rifle (Gould 2004, 189). Leopold (1919) described shooting wild turkeys with his Winchester, which is exhibited at the University of Wisconsin (E. Leopold and S. Temple, pers. comm.). Greene 2011 mistakenly describes Snyder as "hunting" with a model '94, whereas in fact Gary finished off and butchered deer wounded by hunters (pers. comm.).
5. Pruetz and Bertolani 2007.
6. See, e.g., Nelson and Callicott 2008.
7. Hillis 1990, 4.
8. Hutchinson 1965.
9. Haley 1936, 462.
10. Dobie 1941, 33.
11. http://doublehelixranch.com (accessed Jan. 13, 2013).
12. Schaller, pers. comm.
13. McMurtry 1985.
14. Klauber 1956.
15. See Kowalsky 2010 for a recent overview.
16. Clutton-Brock 1999.
17. Faulkner 1942.
18. Tantillo 2001.

CHAPTER FOURTEEN

1. Dioum's quote is on public display at the Bronx Zoo.
2. Herman 1986.
3. Title borrowed from Loveridge 1944.
4. See especially Neruda 1965.
5. Angier 1991.
6. See Klauber 1956 for early mentions of rattlers, including John Smith reference; Franklin's rattlesnake essay from an editorial sidebar in Greene 2003, 30–31.
7. Brown 1993.
8. Brown (1993) summarized earlier work; see also Reinert et al. 2011 and Clark et al. 2012.
9. Greene and Campbell 1992.
10. Fitch 1949.

11. MacArthur and Pianka 1966; Emlen 1966.

12. Fitch et al. 2001.

13. Fitch 1999.

14. Lillywhite and McCleary 2008.

15. Hatteberg 1991, 48.

16. Fraser 2001.

17. For starters, see *The Myth of Sisyphus* (Camus 1955), *Turtle Island* (Snyder 1974), *Legends of the Fall* (Harrison 1978), and *Dream Work* (Oliver 1986). For Oliver reading three poems, see www.youtube.com/watch?v = XnaP7ig69go; for a walk with Snyder and Harrison, get the hardcover/DVD version of Ebenkamp 2010.

18. McMurtry 1985, 628.

19. Turner 1996.

20. Mighetto 1986, xxiii.

21. Peacock 1990; Thomas 1990.

22. Hill and Hurtado 1996; Headland and Greene 2011.

23. Stent 1972.

24. They also stole credit from Rosalind Franklin, as is obvious in Watson's (1968) account of discovering DNA's structure.

25. Boyd 2003.

26. Packer and Clottes 2000, 57.

27. Janzen 1998.

28. Eisner 2009.

29. Fitzpatrick et al. 2005.

30. Min et al. 2005.

31. Angier 2002.

32. Fitch and Pisani 2002.

33. Bozarth et al. 2011.

34. Truett and Phillips 2009.

35. The Smithsonian zoologist was Spencer Fullerton Baird, who in a letter to another naturalist described Kennerly as "the most notorious snake, salamander, bug, cave-bone, wolf, panther, and tadpole catcher in the community, your humble servant perhaps excepted" (Good and Wiedenfeld 1995, 629).

36. Duellman 1960.

BIBLIOGRAPHY

Abbey, E. 1988. *Desert Solitaire.* University of Arizona Press, Tucson.

Amarello, M., J.J. Smith, and J. Sloane. 2011. Family values: rattlesnake parental care is more than just attendance. Oral presentation at the Biology of the Rattlesnakes Symposium, Tucson, AZ.

Anderson, P. 1942. Amphibians and reptiles of Jackson County, Missouri. *Bulletin of the Chicago Academy of Sciences* 6:203–220.

———. 1965. *The Reptiles of Missouri.* University of Missouri Press, Columbia.

Andreadis, P. T., and G. M. Burghardt. 1993. Feeding behavior and an oropharyngeal component of satiety in a two-headed snake. *Physiology and Behavior* 54:649–658.

Angier, N. 1991. Pit viper's life: bizarre, gallant and venomous. *New York Times,* October 15: 5, 9.

———. 2002. Venomous and sublime: the viper tells its tale. *New York Times,* December 10.

Armentano, R. A., and M. Schaer. 2011. Overview and controversies in the medical management of pit viper envenomation in the dog. *Journal of Veterinary Emergency and Clinical Care* 21:461–470.

Attenborough, D. 2008. *Life in Cold Blood.* BBC Worldwide, United Kingdom.

Atz, J. W. 1970. The application of the idea of homology to behavior. Pp. 53–74 in L. R. Aronson, E. Tobach, D. S. Lehrman, and J. S. Rosenblatt (eds.), *Development and Evolution of Behavior: Essays in Memory of T. C. Schneirla.* W. H. Freeman, San Francisco.

Auliya, M. A. 2006. *Taxonomy, Life History and Conservation of Giant Reptiles in West Kalimantan.* Natur und Tier Verlag, Berlin.

Baker, E. W. 1945. *Stocky, Boy of West Texas.* John C. Winston, Philadelphia.

Barker, D. G., S. L. Barten, J. P. Ehrsam, and L. Daddono. 2012. The corrected lengths of two well-known giant pythons and the establishment of a new maximum length record for Burmese pythons, *Python bivittatus. Bulletin of the Chicago Herpetological Society* 47:1–6.

Barlow, C. 2000. *The Ghosts of Evolution: Nonsensical Fruit, Missing Partners, and Other Ecological Anachronisms.* Basic Books, New York.

Barnosky, A. D. 2009. *Heatstroke: Nature in an Age of Global Warming.* Island Press, Washington DC.

Bartecki, U., and E. W. Heymann. 1987. Field observation of snake-mobbing in a group of saddle-back tamarins, *Saguinus fuscicollis nigrifrons. Folia Primatologia* 48:199–202.

Bartlett, R. C., and J. Krieger. 1988. *The Sportsman's Guide to Texas: Hunting and Fishing in the Lone Star State.* Taylor Publishing, Dallas.

Bartlett, R. C., and L. Williamson. 1995. *Saving the Best of Texas: A Partnership Approach to Conservation.* University of Texas Press, Austin.

Beaupre, S. J., and H. W. Greene. 2012. Handling live reptiles: leave your ego at the door. Pp. 130–134 in R. W. McDiarmid, M. S. Foster, C. Guyer, J. W. Gibbons, and N. Chernoff (eds.), *Reptile Biodiversity: Standard Methods for Inventory and Monitoring.* University of California Press, Berkeley.

Beebe, W. 1927. *Pheasant Jungles.* G. P. Putnam's Sons, New York.

Bellairs, A. d'A. 1969. *The Life of Reptiles.* Weidenfeld and Nicolson, London.

Blair, W. F. 1950. The biotic provinces of Texas. *Texas Journal of Science* 2:93–117.

Bock, C. E., and J. H. Bock. 2000. *The View from Bald Hill: Thirty Years in an Arizona Grassland.* University of California Press, Berkeley.

Bonnet, X., R. Shine, and O. Lourdais. 2002. Taxonomic chauvinism. *Trends in Ecology and Evolution* 17:1–3.

Boyd, C. E. 2003. *Rock Art of the Lower Pecos.* Texas A&M University Press, College Station.

Bozarth, C. A., F. Hailer, L. L. Rockwood, C. W. Edwards, and J. E. Maldonado. 2011. Coyote colonization of northern Virginia and admixture with Great Lakes wolves. *Journal of Mammalogy* 92:1070–1080.

Brochu, C. A., J. Njau, R. J. Blumenschine, and L. D. Densmore. 2010. A new horned crocodile from the Plio-Pleistocene hominid sites at Olduvai Gorge, Tanzania. *PLoS One* 5(e9333):1–13.

Brown, W. S. 1993. Biology, status, and management of the timber rattlesnake (*Crotalus horridus*): a guide for conservation. *Society for the Study of Amphibians and Reptiles, Herpetological Circular* 12:1–78.

Burghardt, G. M. 1967. Chemical-cue preferences of inexperienced snakes. *Science* 157:718–721.

———. 1992. Human-bear bonding in research on black bear behavior. Pp. 365–382 in H. Davis and D. Balfour (eds.), *The Inevitable Bond: Examining Scientist-Animal Interactions.* Cambridge University Press, Cambridge.

———. 1997. Amending Tinbergen: a fifth aim for ethology. Pp. 254–276 in R. W. Mitchell, N. S. Thompson, and H. L. Miles (eds.), *Anthropomorphism, Anecdotes, and Animals.* SUNY Press, Albany NY.

———. 2005. *The Genesis of Animal Play: Testing the Limits.* MIT Press, Cambridge MA.

Burghardt, G. M., and H. W. Greene. 1988. Predator stimulation and duration of death feigning in neonate hognose snakes. *Animal Behaviour* 36:1842–1844.

Burghardt, G. M., H. W. Greene, and A. S. Rand. 1977. Social behavior in hatchling green iguanas: life at a reptile rookery. *Science* 195:689–691.

Burghardt, G. M., and H. A. Herzog Jr. 1980. Beyond conspecifics: is Brer Rabbit our brother? *BioScience* 30:763–768.

Burghardt, G. M., R. O. Hietalla, and M. R. Pelton. 1972. Knowledge and attitudes concerning black bears by users of the Great Smokey Mountains National Park. Pp. 255–273 in S. Herrero (ed.), *Bears: Their Biology and Management.* International Union for the Conservation of Nature and Natural Resources, Morges, Switz.

Callicott, J. B. 2011. Postmodern ecological restoration: choosing appropriate spatial and temporal scales. Pp. 317–342 in K. de Laplante, B. Brown, and K. S. Peacock (eds.), *Philosophy of Ecology.* Elsevier, Amsterdam.

Camp, C. L. 1916. *Notes on the Local Distribution and Habits of the Amphibians and Reptiles of Southeastern California in the Vicinity of the Turtle Mountains.* University of California Publications in Zoology 12(17). University of California Press, Berkeley.

Campbell, J. A., and W. W. Lamar. 2004. *The Venomous Reptiles of the Western Hemisphere.* Cornell University Press, Ithaca NY.

Campbell, J., B. Moyers, and B. S. Flowers. 1988. *The Power of Myth.* Doubleday, New York.

Camus, A. 1955. *The Myth of Sisyphus.* Alfred A. Knopf, New York.

Canfield, M. R. (ed.). 2011. *Field Notes on Science and Nature.* Harvard University Press, Cambridge MA.

Carothers, J. H. 1986. An experimental confirmation of morphological adaptation: toe fringes in the sand-dwelling lizard *Uma scoparia. Evolution* 40:871–874.

Casewell, N. R., W. Wüster, F. J. Vonk, R. A. Harrison, and B. G. Fry. 2013. Complex cocktails: the evolutionary novelty of venoms. *Trends in Ecology and Evolution* 28:219–229.

Clark, D. R. 1970. *Ecological Study of the Worm Snake* Carphophis vermis (*Kennicott*). University of Kansas Publications, Museum of Natural History 19(2). University of Kansas, Lawrence.

Clark, R. W., W. S. Brown, R. Stechert, and H. W. Greene. 2012. Cryptic sociality in rattlesnakes (*Crotalus horridus*) detected by kinship analysis. *Biology Letters* 8:523–525.

Clutton-Brock, J. 1999. *A Natural History of Domesticated Mammals.* Cambridge University Press, Cambridge.

Cole, C. J. 1966. Femoral glands in lizards: a review. *Herpetologica* 22:199–206.

Colwell, R. K. 1985. A bite to remember. *Natural History* 94(7):56–63.

Conant, R. 1958. *A Field Guide to Reptiles and Amphibians of the United States and Canada East of the 100th Meridian.* Houghton Mifflin, Boston MA.

Cundall, D. 2009. Viper fangs: functional limitations of extreme teeth. *Physiological and Biochemical Zoology* 82:63–79.

Cundall, D., and H. W. Greene. 2000. Feeding in snakes. Pp. 293–333 in K. Schwenk (ed.), *Feeding: Form, Function, and Evolution in Tetrapod Vertebrates.* Academic Press, San Diego.

Darwin, C. 1859. *On the Origin of Species by Means of Natural Selection or the Preservation of Favoured Races in the Struggle for Life.* John Murray, London.

Davies, A. R., A. Corl, Y. Surget-Groba, and B. Sinervo. 2011. Convergent evolution of kin-based sociality in a lizard. *Proceedings of the Royal Society of London B* 278:1507–1514.

De Kay, J. E. 1842. *Zoology of New-York or the New-York Fauna; Comprising Detailed Descriptions of All the Animals Hitherto Observed within the State of New-York, with Brief Notices of Those Occasionally Found Near Its Borders, and Accompanied by Appropriate Illustrations.* D. Appleton and Wiley & Putnam, Boston MA.

De Lang, R. 2010. The reticulated python (*Broghammerus reticulatus*) and man (*Homo sapiens*) eat each other: animals, enjoy your meal! *Litteratura Serpentium* 30:254–269.

Deufel, A., and D. Cundall. 2006. Functional plasticity of the venom delivery system in snakes with a focus on the poststrike prey release behavior. *Zoologischer Anzeiger* 245:249–267.

Dial, B. E. 1965. Pattern and coloration in juveniles of two West Texas *Elaphe*. *Herpetologica* 21:75–78.

———. 1978. Aspects of the behavioral ecology of two Chihuahuan Desert geckos (Reptilia, Lacertilia, Gekkonidae). *Journal of Herpetology* 12:209–216.

———. 1987a. Energetics and performance during nest emergence and the hatchling frenzy in loggerhead sea turtles (*Caretta caretta*). *Herpetologica* 43:307–315.

———. 1987b. Energetics of concertina locomotion in *Bipes biporus* (Reptilia: Amphisbaenia). *Copeia* 1987:470–477.

Dial, B. E., and L. C. Fitzpatrick. 1983. Lizard tail autotomy: function and energetics of postautotomy tail movement in *Scincella lateralis*. *Science* 219:391–393.

Dial, B. E., and L. L. Grismer. 1992. A phylogenetic analysis of physiological-ecological character evolution in the lizard genus *Coleonyx* and its implications for historical biogeographic reconstruction. *Systematic Biology* 41:178–195.

Ditmars, R. L. 1931. *Snakes of the World.* Macmillan, New York.

———. 1939. *Thrills of a Naturalist's Quest.* Macmillan, New York.

Dixon, J. R., and J. W. Wright. 1975. A review of the iguanid lizard genus *Tropidurus* in Peru. *Contributions in Science, Natural History Museum of Los Angeles County* 271:1–39.

Dobie, J. F. 1941. *The Longhorns.* Little, Brown, Boston.

Donlan, C. J., J. Berger, C. E. Bock, J. H. Bock, D. A. Burney, J. A. Estes, D. Foreman, P. S. Martin, G. W. Roemer, F. A. Smith, M. E. Soulé, and H. W. Greene. 2006. Pleistocene rewilding: an optimistic agenda for 21st century conservation. *American Naturalist* 168:660–681.

Donlan, C. J., and H. W. Greene. 2010. NLIMBY: no lions in my backyard. Pp. 293–305 in M. Hall (ed.), *Restoration and History: The Search for a Usable Environmental Past.* Routledge, London.

Donlan, C. J., H. W. Greene, J. Berger, C. E. Bock, J. H. Bock, D. A. Burney, J. A. Estes, D. Foreman, P. S. Martin, G. W. Roemer, F. A. Smith, and M. E. Soulé. 2005. Re-wilding North America. *Nature* 436:913–914.

Drees, W. B. (ed.). 2003. *Is Nature Ever Evil?* Routledge, London.

Duellman, W. E. 1960. A record size for *Drymarchon corais melanurus. Copeia* 1960:367–368.

———. 2009. Reminiscences of Henry Fitch. *Herpetological Review* 40:395.

Dunn, E. R. 1954. The coral snake mimicry problem. *Evolution* 8:97–102.

Eakin, R. M. 1975. *Great Scientists Speak Again.* University of California Press, Berkeley.

Earthworks Group. 1989. *50 Simple Things You Can Do to Save the Earth.* Earthworks Press, Berkeley CA.

Ebenkamp, P. 2010. *The Etiquette of Freedom: Gary Snyder, Jim Harrison, and the Practice of the Wild.* Counterpoint Press, Berkeley CA.

Eibl-Eibesfeldt, I. 1970. *Ethology: The Biology of Behavior.* Holt, Rinehart and Winston, New York.

Eisner, T. 2009. *Eisner's World: Life through Many Lenses.* Sinauer Associates, Sunderland MA.

Emlen, J. M. 1966. The role of time and energy in food preference. *American Naturalist* 100:611–617.

Ewer, R. F. 1968. *Ethology of Mammals.* Logos Press, London.

Faulkner, W. 1942. *"Go Down, Moses" and Other Stories.* Random House, New York.

Finkelstein, M. E., D. F. Doak, D. George, J. Burnett, J. Brandt, M. Church, J. Grantham, and D. R. Smith. 2012. Lead poisoning and the deceptive recovery of the critically endangered California condor. *Proceedings of the National Academy of Sciences* 109:11449–11454.

Fitch, H. S. 1935. *Natural History of the Alligator Lizards.* Transactions of the Academy of Science of Saint Louis 29(1). Academy of Science, St. Louis.

———. 1940. *A Biogeographical Study of the Ordinoides Artenkreis of Garter Snakes (Genus* Thamnophis*).* University of California Publications in Zoology 44(1). University of California Press, Berkeley.

———. 1948a. Further remarks concerning *Thamnophis ordinoides* and its relatives. *Copeia* 1948:121–126.

———. 1948b Habits and economic relationships of the Tulare kangaroo rat. *Journal of Mammalogy* 29:5–35.

———. 1948c. Ecology of the California ground squirrel on grazing lands. *American Midland Naturalist* 39:513–596.

———. 1949. Outline for ecological life history studies of reptiles. *Ecology* 30:520–532.

———. 1951. A simplified type of funnel trap for reptiles. *Herpetologica* 7:77–80.

———. 1958. *Home Ranges, Territories, and Seasonal Movements of Vertebrates of the Natural History Reservation.* University of Kansas Publications, Museum of Natural History 11(3). University of Kansas, Lawrence.

———. 1960a. *Autecology of the Copperhead.* University of Kansas Publications, Museum of Natural History 13(4). University of Kansas, Lawrence.

———. 1960b. Criteria for determining sex and breeding maturity in snakes. *Herpetologica* 16:49–51.

————. 1961. The snake as a source of living spermatozoa in the laboratory. *Turtox News* 39:247.

————. 1963. *Spiders of the University of Kansas Natural History Reservation and Rockefeller Experimental Tract*. University of Kansas Museum of Natural History, Miscellaneous Publications 33. University of Kansas, Lawrence.

————. 1989. A *Field Study of the Slender Glass Lizard*, Ophisaurus attenuatus, *in Northeastern Kansas*. Occasional Papers of the Museum of Natural History, University of Kansas 125. University of Kansas, Lawrence.

————. 1999. *A Kansas Snake Community: Composition and Changes over 50 Years*. Krieger, Malabar FL.

Fitch, H. S., and A. F. Echelle. 2000. Historical perspective: Henry S. Fitch. *Copeia* 2000:891–900.

Fitch, H. S., and V. R. Fitch. 1955. Observations on the summer tanager in northeastern Kansas. *Wilson Bulletin* 67:45–54.

Fitch, H. S., and H. W. Greene. 1965. *Breeding Cycle in the Ground Skink*, Lygosoma laterale. Publications of the Museum of Natural History, University of Kansas 15(11). University of Kansas, Lawrence.

Fitch, H. S., and R. W. Henderson. 1973. A new anole (Reptilia: Iguanidae) from southern Veracruz, Mexico. *Journal of Herpetology* 7:125–128.

Fitch, H. S., and G. R. Pisani. 2002. Longtime recapture of a timber rattlesnake (*Crotalus horridus*) in Kansas. *Journal of the Kansas Herpetological Society* 3:15–16.

Fitch, H. S., G. R. Pisani, H. W. Greene, A. F. Echelle, and M. Zerwekh. 2004. A field study of the timber rattlesnake in Leavenworth County, Kansas. *Journal of the Kansas Herpetological Society* 11:18–24.

Fitch, H. S., P. von Achen, and A. F. Echelle. 2001. A half-century of forest invasion on a natural area in northeastern Kansas. *Transactions of the Kansas Academy of Sciences* 104:1–17.

Fitch, H. S., P. von Achen, and G. L. Pittman. 2003. Probable succession-related prey changes of long-eared owls in Kansas. *Kansas Ornithological Society Bulletin* 54:42–43.

Fitch, V. R. 1984. The published contributions of Henry S. Fitch. Pp. 5–9 in R. A. Seigel et al. (eds.), *Vertebrate Ecology and Systematics: A Tribute to Henry S. Fitch*. University of Kansas Museum of Natural History Special Publication 10. University of Kansas, Lawrence.

Fitzpatrick, J. W., M. Lammertink, M. D. Luneau, T. W. Gallagher, B. R. Harrison, G. M. Sparling, K. V. Rosenberg, R. W. Rohrbaugh, E. C. H. Swarthout, P. H. Wrege, S. B. Swarthout, M. S. Dantzker, R. A. Charif, T. R. Barksdale, J. V. Remsen Jr., S. D. Simon, and D. Zollner. 2005. Ivory-billed woodpecker (*Campephilus principalis*) persists in North America. *Science* 308:1460–1462.

Flannery, T. 2002. A new Darwinism? Review of S. J. Gould, *The Structure of Evolutionary Theory* and *I Have Landed: The End of a Beginning in Natural History*. *New York Review of Books,* May 23: 52–54.

Fleischner, T. (ed.). 2011. *The Way of Natural History*. Trinity University Press, San Antonio TX.

Fraser, C. 2001. Pretty in sunlight. *New York Review of Books,* Oct. 4: 41–43.

———. 2009. *Rewilding the World: Dispatches from the Conservation Revolution.* Metropolitan Books, New York.

Fredriksson, G. M. 2005. Predation on sun bears by reticulated python in East Kalimantan, Indonesian Borneo. *Raffles Bulletin of Zoology* 53:165–168.

French, M. 1977. *The Women's Room: A Novel.* Jove Books, New York.

Gans, C. 1986. Locomotion of limbless vertebrates: patterns and evolution. *Herpetologica* 42:33–46.

Garcia Márquez, G. 1967. *One Hundred Years of Solitude.* Harper and Row, New York.

Gehlbach, F. R. 1972. Coral snake mimicry reconsidered: the strategy of self-mimicry. *Forma et Functio* 5:311–320.

———. 1981. *Mountain Islands and Desert Seas: A Natural History of the U.S.-Mexican Borderlands.* Texas A&M University Press, College Station.

Gehlbach, F. R., J. F. Watkins II, and J. C. Kroll. 1971. Pheromone trail-following studies of typhlopid, leptotyphlopid, and colubrid snakes. *Behaviour* 40: 282–294.

Goin, C. J., O. B. Goin, and G. R. Zug. 1978. *Introduction to Herpetology.* W. H. Freeman, San Francisco CA.

Good, D. A., and M. G. Wiedenfeld. 1995. The holotype of the Texas alligator lizard, *Gerrhonotus infernalis* Baird (Squamata: Anguidae). *Journal of Herpetology* 29:628–630.

Goris, R. C. 2011. Infrared organs of snakes: an integral part of vision. *Journal of Herpetology* 45:2–14.

Gould, C. 2004. *The Remarkable Life of William Beebe: Explorer and Naturalist.* Island Press, Washington DC.

Gould, S. J. 1977. *Ontogeny and Phylogeny.* Harvard University Press, Cambridge MA.

———. 1983. A life and death tail. *Natural History* 92(6):12, 14–16.

Gould, S. J., and R. C. Lewontin. 1979. The Spandrels of San Marco and the Panglossian paradigm: a critique of the adaptationist programme. *Proceedings of the Royal Society of London B* 205:581–598.

Gould, S. J., and E. Vrba. 1982. Exaptation—a missing term in the science of form. *Paleobiology* 8:4–15.

Greene, H. W. 1961. Additional instances of tail-waving in salamanders. *Bulletin of the Philadelphia Herpetological Society* 9:19.

———. 1969a. Reproduction in a Middle American skink, *Leiolopisma cherriei* (Cope). *Herpetologica* 25:55–56.

———. 1969b. Fat storage in females of an introduced lizard, *Hemidactylus turcicus,* in Texas. *Texas Journal of Science* 21:233–235.

———. 1969c. Unusual pattern and coloration in snakes of the genus *Pliocercus* from Veracruz. *Journal of Herpetology* 3:27–31.

———. 1973. Defensive tail display by snakes and amphisbaenians. *Journal of Herpetology* 7:143–161.

———. 1976. Scale overlap, a directional sign stimulus for prey ingestion by ophiophagous snakes. *Zeitschrift für Tierpsychologie* 41:113–120.

———. 1979. Behavioral convergence in the defensive displays of snakes. *Experientia* 35:747–748.

———. 1982. Dietary and phenotypic diversity in lizards: why are some organisms specialized? Pp. 107–128 in D. Mossakowski and G. Roth (eds.), *Environmental Adaptation and Evolution: A Theoretical and Empirical Approach.* G. Fischer-Verlag, Stuttgart.

———. 1983a. Dietary correlates of the origin and radiation of snakes. *American Zoologist* 23:431–441.

———. 1983b. *Boa constrictor* (Boa, Bequer, Boa Constrictor). Pp. 380–382 in D. H. Janzen (ed.), *Costa Rican Natural History.* University of Chicago Press, Chicago.

———. 1984. Feeding behavior and diet of the eastern coral snake, *Micrurus fulvius.* Pp. 147–162 in R. A. Seigel et al. (eds.), *Vertebrate Ecology and Systematics: A Tribute to Henry S. Fitch.* University of Kansas Museum of Natural History Special Publication 10. University of Kansas, Lawrence.

———. 1986. Diet and arboreality in the emerald monitor, *Varanus prasinus,* with comments on the study of adapation. *Fieldiana, Zoology,* n.s. 31:1–12.

———. 1988a. Species richness in tropical predators. Pp. 259–280 in F. Almeda and C. M. Pringle (eds.), *Tropical Rainforests: Diversity and Conservation.* California Academy of Sciences and Pacific Division, American Association for the Advancement of Science, San Francisco.

———. 1988b. Antipredator mechanisms in reptiles. Pp. 1–152 in C. Gans and R. B. Huey (eds.), *Biology of the Reptilia,* vol. 16, *Ecology B: Defense and Life History.* Alan R. Liss, New York.

———. 1989. Agonistic behavior by three-toed sloths, *Bradypus variegatus.* *Biotropica* 21:369–372.

———. 1994. Homology and behavioral repertoires. Pp. 369–391 in B. K. Hall (ed.), *Homology: The Hierarchical Basis of Comparative Biology.* Academic Press, San Diego.

———. 1997. *Snakes: The Evolution of Mystery in Nature.* University of California Press, Berkeley.

———. 1999a. Natural history and behavioural homology. Pp. 173–188 in G. R. Bock and G. Cardew (eds.), *Homology,* Novartis Foundation Symposium 222. John Wiley and Sons, Chichester, U.K.

———. 1999b. Benjamin Edward Dial (1944–1998). *Herpetological Review* 30:6–7.

———. 2003. Appreciating rattlesnakes. *Wild Earth* 13(2/3):28–32.

———. 2005. Organisms in nature as a central focus for biology. *Trends in Ecology and Evolution* 20:23–27.

———. 2009. Henry Fitch and the practice of natural history. *Herpetological Review* 40:393–394.

———. 2010. Anaconda! *Science* 327:1577.

————. 2011. Longhorns, whitetails, and the evolution of "wild." Pp. 311–321 in J. B. Losos (ed.), *In the Light of Evolution: Essays from the Laboratory and Field*. Roberts and Co., Greenwood Village CO.

Greene, H. W., and G. M. Burghardt. 1978. Behavior and phylogeny: constriction in ancient and modern snakes. *Science* 200:74–77.

Greene, H. W., G. M. Burghardt, B. A. Dugan, and A. S. Rand. 1978. Predation and the defensive behavior of green iguanas (Reptilia, Lacertilia, Iguanidae). *Journal of Herpetology* 12:169–176.

Greene, H. W., and J. A. Campbell. 1992. The future of pitvipers. Pp. 421–427 in J. A. Campbell and E. D. Brodie Jr. (eds.), *Biology of the Pitivipers*. Selva, Tyler TX.

Greene, H. W., and B. E. Dial. 1966. Brooding behavior by female Texas alligator lizards. *Herpetologica* 22:303.

Greene, H. W., and F. M. Jaksic. 1983. Food niche relationships among sympatric predators: effects of level of prey identification. *Oikos* 40:151–154.

————. 1992. The feeding behavior and natural history of two Chilean snakes, *Philodryas chamissonis* and *Tachymenis chilensis* (Colubridae). *Revista Chilena de Historia Natural* 65:485–493.

Greene, H. W., P. May, D. L. Hardy Sr., J. Sciturro, and T. Farrell. 2002. Parental behavior by vipers. Pp. 179–205 in G. W. Schuett, M. Höggren, M. E. Douglas, and H. W. Greene (eds.), *Biology of the Vipers*. Eagle Mountain Publishing, Eagle Mountain UT.

Greene, H. W., and R. W. McDiarmid. 2005. Wallace and Savage: heroes, theories, and venomous snake mimicry. Pp. 190–208 in M. A. Donnelly, B. I. Crother, C. E. Guyer, M. H. Wake, and M. E. White (eds.), *Ecology and Evolution in the Tropics: A Herpetological Perspective*. University of Chicago Press, Chicago.

Greene, H. W., and G. V. Oliver Jr. 1965. Notes on the natural history of the western massasauga. *Herpetologica* 21:225–228.

Greene, H. W., and W. F. Pyburn. 1973. Comments on aposematism and mimicry among coral snakes. *The Biologist* 55:144–148.

Greene, H. W., P. M. Ralidis, and E. W. Acuña. 2009. Texas alligator lizard. Pp. 492–495 in L. L. C. Jones and R. Lovich (eds.), *Lizards of the American Southwest: A Photographic Field Guide*. Rio Nuevo Publishers, Tucson AZ.

Greene, H. W., and M. J. Wakeman. 1962. Ozark herpetozoa in Johnson County, Missouri. *Herpetologica* 18:127.

Griffin, D. R. 1976. *The Question of Animal Awareness: Evolutionary Continuity of Mental Experience*. Rockefeller University Press, New York.

Haju, D. 2001. *Positively 4th Street: The Lives and Times of Joan Baez, Bob Dylan, Mimi Baez Fariña, and Richard Fariña*. Farrar, Straus, and Giroux, New York.

Halberstam, D. 1972. *The Best and the Brightest*. Ballantine Books, New York.

Haley, J. E. 1936. *Charles Goodnight: Cowman and Plainsman*. University of Oklahoma Press, Norman.

Hardy, D. L., Sr., and K. R. Zamudio. 2006. Compartment syndrome, fasciotomy, and neuropathy after a rattlesnake envenomation: aspects of monitoring and diagnosis. *Wilderness and Environmental Medicine* 17:36–40.

Harrison, J. 1978. *Legends of the Fall.* Delacorte Press, New York.

Hatteberg, L. 1991. *Larry Hatteberg's Kansas People: A Collection of Colorful Personalities from the Sunflower State.* Wichita Eagle and Beacon Publishing, Wichita KS.

Haverschmidt, F. 1970. Wattled jacana caught by an anaconda. *Condor* 72:364.

Head, J. J., J. I. Bloch, A. K. Hastings, J. R. Bourque, E. A. Cadena, F. A. Herrera, P. D. Polly, and C. A. Jaramillo. 2009. Giant boid snake from the Paleocene neotropics reveals hotter past equatorial temperatures. *Nature* 457:715–718.

Headland, T. N., and H. W. Greene. 2011. Hunter-gatherers and other primates as prey, predators, and competitors of snakes. *Proceedings of the National Academy of Sciences* 108:20865–20866, E1470–E1474.

Hedges, S. B., W. E. Duellman, and M. P. Heinicke. 2008. New World direct-developing frogs (Anura: Terrarana): molecular phylogeny, classification, biogeography, and conservation. *Zootaxa* 1737:1–182.

Henderson, R. W. 2009. Henry Fitch at home and in the tropics. *Herpetological Review* 40:396.

Herman, S. G. 1986. *The Naturalist's Field Journal: A Manual of Instruction Based on a System Established by Joseph Grinnell.* Buteo Books, Vermillion SD.

Herzog, H. 2010. *Some We Love, Some We Hate, Some We Eat: Why It's So Hard to Think Straight about Animals.* Harper Collins, New York.

Heymann, E. W. 1987. A field observation of predation on a mustached tamarin (*Saguinus mystax*) by an anaconda. *International Journal of Primatology* 8:193–195.

Hill, K., and A. M. Hurtado. 1996. *Aché Life History: The Ecology and Demography of a Foraging People.* Aldine de Gruyter, New York.

Hillis, D. M. 1990. *A New Species of Xenodontine Colubrid Snake of the Genus Synophis from Ecuador and the Phylogeny of the Genera Synophis and Emmochliophis.* Occasional Papers of the Museum of Natural History, University of Kansas 135. University of Kansas, Lawrence.

———. 2009. In the field with Henry Fitch. *Herpetological Review* 40:396–398.

Holmberg, A. R. 1957. Lizard hunts on the north coast of Peru. *Fieldiana: Anthropology* 36:203–220.

Hubbs, C. 1941. Reviews and comments: A biogeographical study of the *ordinoides* artenkreis of garter snakes (genus *Thamnophis*). *American Naturalist* 75: 384–386.

Hutchinson, D. A., A. Mori, A. H. Savitzky, G. M. Burghardt, X. Wu, J. Meinwald, and F. C. Schroeder. 2007. Dietary sequestration of defensive steroids in nuchal glands of the Asian snake *Rhabdophis tigrinus*. *Proceedings of the National Academy of Sciences* 104:2265–2270.

Hutchinson, G. E. 1965. *The Ecological Theater and the Evolutionary Play.* Yale University Press, New Haven CT.

Irving, J. 1981. *The Hotel New Hampshire.* E. P. Dutton, New York.

Isbell, L. A. 2009. *The Fruit, the Tree, and the Serpent: Why We See Well.* Harvard University Press, Cambridge MA.

Jaksic, F. M., and H. W. Greene. 1984. Empirical evidence of non-correlation between tail loss frequency and predation intensity on lizards. *Oikos* 42:407–411.

Jaksic, F. M., H. W. Greene, and J. L. Yañez. 1981. The guild structure of a community of predatory vertebrates in central Chile. *Oecologia* 49:21–28.

Janzen, D. H. 1977. Why fruit rots, seeds mold, and meat spoils. *American Naturalist* 111:691–713.

——— (ed.). 1983. *Costa Rican Natural History.* University of Chicago Press, Chicago.

———. 1998. Gardenification of wildland nature and the human footprint. *Science* 279:1312–1313.

Janzen, D. H., and P. S. Martin. 1982. Neotropical anachronisms: the fruits the gomphotheres ate. *Science* 215:19–27.

Jeffers, R. 1988. *The Collected Poetry of Robinson Jeffers.* Stanford University Press, Stanford CA.

Jones, L. L. C., and R. Lovich (eds.). 2009. *Lizards of the American Southwest: A Photographic Field Guide.* Rio Nuevo Publishers, Tucson AZ.

Kareiva, P., and M. Marvier. 2012. What is conservation science? *BioScience* 62:962–969.

Kay, C., and R. T. Simmons. 2002. *Wilderness and Political Ecology: Aboriginal Influences and the Original State of Nature.* University of Utah Press, Salt Lake City.

Kellert, S. R., and E. O. Wilson. 1993. *The Biophilia Hypothesis.* Island Press, Washington DC.

Kiester, A. R. 1997. Aesthetics of biological diversity. *Human Ecology Review* 3:151–163.

Klauber, L. M. 1956. *Rattlesnakes: Their Habits, Life Histories, and Influence on Mankind.* University of California Press, Berkeley.

Klopfer, P. H. 1962. *Behavioral Aspects of Ecology.* Prentice-Hall, Englewood Cliffs NJ.

———. 1973. Does behavior evolve? *Annals of the New York Academy of Sciences* 223:113–125.

———. 1999. *Politics and People in Ethology: Personal Reflections on the Study of Animal Behavior.* Associated University Presses, Cranbury NJ.

Kowalsky, N. (ed). 2010. *Hunting: In Search of the Wild Life.* Wiley-Blackwell, West Sussex, U.K.

Kramer, E. 1977. Zur Schlangenfauna Nepals. *Revue Suisse de Zoologie* 84:721–761.

Kuhn, T. S. 1962. *The Structure of Scientific Revolutions.* University of Chicago Press, Chicago.

Kurtén, B., and E. Anderson. 1980. *Pleistocene Mammals of North America.* Columbia University Press, New York.

Larrick, J. W., J. A. Yost, and J. Kaplan. 1978. Snake bite among the Waorani Indians of eastern Ecuador. *Transactions of the Royal Society of Tropical Medicine and Hygiene* 72:542–543.

Larson, A., and J. B. Losos. 1996. Phylogenetic systematics of adapatation. Pp. 187–220 in M. R. Rose and G. V. Lauder (eds.), *Adaptation*. Academic Press, San Diego CA.

Lee, J. 1996. *The Amphibians and Reptiles of the Yucatan Peninsula*. Cornell University Press, Ithaca NY.

Leopold, A. S. 1919. A turkey hunt in the Datil National Forest. *Wild Life,* Dec.: 4–5, 16.

Levy, S. 2012. *Once and Future Giants: What Ice Age Extinctions Tell Us about the Fate of Earth's Largest Animals*. Oxford University Press, New York.

Lillywhite, H. B., and R. W. Henderson. 1993. Behavioral and functional ecology of arboreal snakes. Pp. 1–48 in R. A. Seigel and J. T. Collins (eds.), *Snakes: Ecology and Behavior*. McGraw-Hill, New York.

Lillywhite, H. B., and R. J. R. McCleary. 2008. Trophic ecology of insular cottonmouth snakes: review and perspective. *South American Journal of Herpetology* 3:175–185.

Losos, J. B., and H. W. Greene 1988. Ecological and evolutionary implications of diet in monitor lizards. *Biological Journal of the Linnean Society* 35:379–407.

Loveridge, A. 1944. *Many Happy Days I've Squandered*. Harper and Brothers, New York.

Luke, C. 1986. Convergent evolution of lizard toe fringes. *Biological Journal of the Linnean Society* 27:1–16.

MacArthur, R. H. 1972. *Geographical Ecology: Patterns in the Distribution of Species*. Harper and Row, New York.

MacArthur, R. H., and E. R. Pianka. 1966. On the optimal use of a patchy environment. *American Naturalist* 100:603–609.

Mackessy, S. P. 2010. Evolutionary trends in venom composition in the western rattlesnake (*Crotalus viridis* sensu lato): toxicity versus tenderizers. *Toxicon* 55: 1463–1474.

Manning, A. D., D. B. Lindenmayer, and J. Fischer. 2006. Stretch goals and backcasting: approaches for overcoming barriers to large-scale ecological restoration. *Restoration Ecology* 14:487–492.

Marešová, J., E. Landová, and D. Frynta. 2009. What makes some species of milk snakes more attractive to humans than others? *Theory in Biosciences* 128: 227–235.

Marris, E. 2011. *Rambunctious Garden: Saving Nature in a Post-wild World*. Bloomsbury Press, New York.

Martin, P. S. 1969. Wanted: a suitable herbivore. *Natural History,* Feb.: 35–39.

———. 2005. *Twilight of the Mammoths: Ice Age Extinctions and the Rewilding of America*. University of California Press, Berkeley.

Martin, P. S., and D. Burney. 1999. Bring back the elephants! *Wild Earth* 9:57–64.

Matthiessen, P. 2008. *Shadow Country*. Random House, New York.

Mayr, E. 1941. What is an artenkreis? *Copeia* 1941:115–116.

———. 1942. *Systematics and the Origin of Species*. Columbia University Press, New York.

————. 1983. How to carry out the adaptationist program? *American Naturalist* 121:324–334.

McMurtry, L. 1985. *Lonesome Dove*. Simon and Schuster, New York.

Merker, G., and W. Merker. 2005. *Alterna: The Gray-Banded Kingsnake*. LM Digital, Benicia CA.

Mighetto, L. (ed.) 1986. *Muir among the Animals: The Wildlife Writings of John Muir*. Sierra Club Books, San Francisco.

Miller, A. H. 1964. Joseph Grinnell. *Systematic Zoology* 13:235–242.

Min, M. S., S. Y. Yang, R. M. Bonett, D. R. Vietes, R. A. Brandon, and D. B. Wake. 2005. Discovery of the first Asian plethodontid salamander. *Nature* 435:87–90.

Minton, S. A. 1966. *A Contribution to the Herpetology of West Pakistan*. Bulletin of the American Museum of Natural History 134(2).

Moir, J. 2006. *Return of the Condor: The Race to Save Our Largest Bird from Extinction*. Lyons Press, Guilford CT.

Moffett, M. 2002. Bit. *Outside,* April, 102–105.

Montgomery, G. G., and A. S. Rand. 1978. Movements, body temperature, and hunting strategy of a *Boa constrictor*. *Copeia* 1978:532–533.

Mulcahy, D. G. 2008. Phylogeography and species boundaries of the western North American nightsnake (*Hypsiglena torquata*): revisiting the subspecies concept. *Molecular Phylogenetics and Evolution* 46:1095–1115.

Murphy, J. C., and R. W. Henderson. 1997. *Tales of Giant Snakes: A Historical Natural History of Anacondas and Pythons*. Krieger, Malabar FL.

Myers, G. S. 1941. [Review of] A biogeographical study of the *ordinoides* artenkreis of garter snakes (genus *Thamnophis*). *Copeia* 1941:122–123.

Nabhan, G. P. 2003. *Singing the Turtles to Sea: The Comcac (Seri) Art and Science of Reptiles*. University of California Press, Berkeley.

Nelson, M. P., and J. B. Callicott (eds.). 2008. *The Wilderness Debate Rages On: Continuing the Great New Wilderness Debate*. University of Georgia Press, Athens.

Neruda, P. 1965. *Bestiary/Bestiario*. Harcourt, Brace and World, New York.

Nisbett, A. 1976. *Konrad Lorenz: A Biography*. Harcourt Brace Jovanovich, New York.

Njau, J. K., and R. Blumenschine. 2011. Crocodylian and mammalian carnivore feeding traces on hominid fossils from FLK 22 and FLK NN 3, Plio-Pleistocene, Oluduvai Gorge, Tanzania. *Journal of Human Evolution* 63:408–417.

Noss, R. F. 2012. *Forgotten Grasslands of the South: Natural History and Conservation*. Island Press, Washington DC.

Oliver, J. A. 1955. *The Natural History of North American Amphibians and Reptiles*. D. Van Nostrand, New York.

Oliver, M. 1986. *Dream Work*. Atlantic Monthly Press, Boston.

Packer, C., and J. Clottes. 2000. When lions ruled France. *Natural History* 109(9):52–57.

Packer, C., D. Ikanda, B. Kissui, and H. Kushnir. 2005. Lion attacks on humans in Tanzania. *Nature* 436:927–928.

Peacock, D. 1990. *Grizzly Years: In Search of the American Wilderness.* Henry Holt, New York.

Phelps, T. 2010. *Old World Vipers: A Natural History of the Azemiopinae and Viperinae.* Edition Chimaira, Frankfurt am Main.

Picado, C. T. 1931. *Serpientes venenosas de Costa Rica.* Imprinta Alsina, San José, Costa Rica.

Pisani, G. H. 2009. Henry Fitch: the twilight of an incredible career. *Herpetological Review* 40:399–400.

Pitman, C. R. S. 1974. *A Guide to the Snakes of Uganda.* Revised edition. Weldon & Wesley, Codicote, U.K.

Plummer, M. V. 1977. Reproduction and growth in the turtle *Trionyx muticus. Copeia* 1977:440–447.

———. 2009. Memories of Henry Fitch. *Herpetological Review* 40:395.

Pope, C. H. 1957. Reptiles round the world. Alfred A. Knopf, New York.

Popper, K. 1963. *Conjectures and Refutations: The Growth of Scientific Knowledge.* Routledge, London.

Poran, N. S., R. G. Coss, and E. Benjamini. 1987. Resistance of California ground squirrels (*Spermophilus beecheyi*) to the venom of the northern Pacific rattlesnake (*Crotalus viridis oreganus*): a study of adaptive variation. *Toxicon* 25: 767–778.

Pramuk, J. (ed.). 2008. The reptile training and enrichment issue! *Herp Herald,* Fall/Winter: 1–6.

Preutz, J. D., and P. Bertolani. 2007. Savanna chimpanzees, *Pan troglodytes verus,* hunt with tools. *Current Biology* 17:412–417.

Pringle, C. M., I. Chacon, M. H. Grayum, H. W. Greene, G. S. Hartshorn, G. E. Schatz, F. G. Stiles, C. Gomez, and M. Rodriguez. 1984. Natural history observations and ecological evaluation of the La Selva Protection Zone, Costa Rica. *Brenesia* 22:189–206.

Pyburn, W. F. 1955. Species discrimination in two sympatric lizards, *Sceloporus olivaceus* and *S. poinsetti. Texas Journal of Science* 7:312–315.

———. 1961. The inheritance and distribution of vertebral stripe color in the cricket frog. Pp. 235–261 in W. F. Blair (ed.), *Vertebrate Speciation: A University of Texas Symposium.* University of Texas Press, Austin.

———. 1963. Observations on the life history of the treefrog *Phyllomedusa callidryas* (Cope). *Texas Journal of Science* 15:155–170.

Pyron, R. A., and F. T. Burbrink. 2012. Extinction, ecological opportunity, and the origins of global snake diversity. *Evolution* 66: 163–178.

Rage, J.-C., and S. Bailon. 2011. Amphibia and Squamata. Pp. 467–478 in T. Harrison (ed.), *Paleontology and Geology of Laetoli: Human Evolution in Context.* Springer Verlag, New York.

Rand, A. S. 1968. A nesting aggregation of iguanas. *Copeia* 1968:552–561.

Rand, A. S., and H. W. Greene. 1982. A latitudinal trend in the timing of iguana nesting. Pp. 142–149 in G. M. Burghardt and A. S. Rand (eds.), *Iguanas of the World: Behavior, Ecology and Evolution.* Noyes Publications, Park Ridge NJ.

Regal, P. J. 1966. Thermophilic responses following feeding in certain reptiles. *Copeia* 1966:588–590.

Reilly, S. M., L. B. McBrayer, and D. B. Miles. 2007. *Lizard Ecology: The Evolutionary Consequences of Foraging Mode.* Cambridge University Press, Cambridge.

Reinert, H. K., and D. Cundall. 1982. An improved surgical implantation method for radio-tracking snakes. *Copeia* 1982:702–705.

Reinert, H. K., G. A. MacGregor, L. M. Bushar, and R. T. Zappalorti. 2011. Foraging ecology of timber rattlesnakes, *Crotalus horridus. Copeia* 2011:430–442.

Rivas, J. A. 1998. Predatory attack of a green anaconda (*Eunectes murinus*) on an adult human. *Herpetological Natural History* 6:158–160.

Rivas, J. A., and G. M. Burghardt. 2001. Understanding sexual size dimorphism in snakes: wearing the snake's shoes. *Animal Behaviour* 62:1–6.

Rivas, J. A., M. C. Muñoz, J. B. Thorbjarnarson, G. M. Burghardt, W. Holmstrom, and P. Calle. 2007. Natural history of the green anaconda in the Venezuelan llanos. Pp. 128–138 in R. W. Henderson and R. Powell (eds.), *Biology of the Boas and Pythons.* Eagle Mountain Publishing, Eagle Mountain UT.

Robbins, T. 1980. *Still Life with Woodpecker.* Bantam Books, New York.

Rodríguez-Robles, J. A., D. A. Good, and D. B. Wake. 2003. *Brief History of Herpetology in the Museum of Vertebrate Zoology, University of California, Berkeley, with a List of Type Specimens of Recent Amphibians and Reptiles.* University of California Publications in Zoology 131. University of California Press, Berkeley.

Rodríguez-Robles, J. A., D. G. Mulcahy, and H. W. Greene. 1999. Feeding ecology of the desert nightsnake, *Hypsiglena torquata* (Colubridae). *Copeia* 1999:93–100.

Ross, C. A. 1989. *Crocodiles and Alligators.* Facts on File, New York.

Rossman, D. A., N. B. Ford, and R. A. Seigel. 1996. *The Garter Snakes: Evolution and Ecology.* University of Oklahoma Press, Norman.

Rubenstein, D. R., D. I. Rubenstein, P. W. Sherman, and T. A. Gavin. 2006. Pleistocene park: does re-wilding North America represent sound conservation for the 21st century? *Biological Conservation* 132:232–238.

Rubenstein, D., P. Sherman, D. Rubenstein, and T. Caro. 2007. Rewilding rebuttal. *Scientific American,* Oct.: 12.

Ruthven, A. 1908. Variations and genetic relationships of the garter snakes. *Bulletin of the U.S. National Museum* 61:1–201.

Sandler, R. 2012. *The Ethics of Species.* Cambridge University Press, Cambridge.

Savage, J. M. 2002. *The Amphibians and Reptiles of Costa Rica: A Herpetofauna between Two Continents, between Two Seas.* University of Chicago Press, Chicago.

Saviola, A. J., D. Chiszar, C. Busch, and S. P Mackessy. 2013. Molecular basis for prey location in viperid snakes. *BMC Biology* 11:20, doi:10.186/1741–7007–11–20.

Savitzky, A. H. 1980. The role of venom delivery in snake evolution. *Evolution* 34:1194–1204.

Schaller, G. B. 1972. *The Serengeti Lion: A Study of Predator-Prey Relations.* University of Chicago Press, Chicago.

Schmidt, K. P. 1932. Stomach contents of some American coral snakes, with the description of a new species of *Geophis. Copeia* 1932:6–9.

Schwenk, K. 2008. Comparative anatomy and physiology of chemical senses in nonavian aquatic reptiles. Pp. 65–81 in J. G. M. Thewissen and S. Nummela (eds.), *Sensory Evolution on the Threshold: Adaptations in Secondarily Aquatic Vertebrates.* University of California Press, Berkeley.

Seigel, R. A., L. E. Hunt, J. L. Knight, L. Malaret, and N. L. Zuschlag (eds.). 1984. *Vertebrate Ecology and Systematics: A Tribute to Henry S. Fitch.* University of Kansas Museum of Natural History Special Publication 10. University of Kansas, Lawrence.

―――. 2009. Henry Fitch as a mentor and teacher. *Herpetological Review* 40:398–399.

Shine, R. 1977. Habitats, diet, and sympatry in snakes: a study from Australia. *Canadian Journal of Zoology* 55:118–128.

―――. 1991. *Australian Snakes: A Natural History.* Reed Books, Sydney.

Shine, R., P. S. Harlow, J. S. Keogh, and Boeadi. 1998. The influence of sex and body size on food habits of a giant tropical snake, *Python reticulatus. Functional Ecology* 12:248–258.

Skaroff, M. L. 1971. Snakes and Dr. Schweitzer. *Bulletin of the Philadelphia Herpetological Society* 19:25–27.

Skutch, A. F. 1980. *A Naturalist on a Tropical Farm.* University of California Press, Berkeley.

―――. 1998. Biocompatibility: a criterion for conservation. *Revista de Biología Tropical* 46:481–486.

Smith, H. M. 1946. *Handbook of Lizards: Lizards of the United States and of Canada.* Comstock, Ithaca NY.

Smith, P. W. 1964. [Review of] Natural history of the racer *Coluber constrictor. Copeia* 1964:460.

Snyder, G. 1974. *Turtle Island.* New Directions, New York.

―――. 1990. *The Practice of the Wild: Essays.* North Point Press, San Francisco.

Soulé, M. E. 1980. Thresholds for survival: maintaining fitness and evolutionary potential. Pp. 151–169 in M. E. Soulé and B. A. Wilcox (eds.), *Conservation Biology: An Evolutionary-Ecological Perspective.* Sinauer Associates, Sunderland MA.

Soulé, M. E., and G. Lease (eds.). 1995. *Reinventing Nature: Response to Postmodern Deconstruction.* Island Press, Washington DC.

Spawls, S., and B. Branch. 1995. *The Dangerous Snakes of Africa: Natural History, Species Directory, Venoms, and Snakebite.* Ralph Curtis Books, Sanibel Island FL.

Stebbins, R. C. 1954. *Amphibians and Reptiles of Western North America.* McGraw-Hill, New York.

―――. 1966. *A Field Guide to Western Amphibians and Reptiles.* Houghton Mifflin, Boston.

Stein, B. R. 2001. *On Her Own Terms: Annie Montague Alexander and the Rise of Science in the American West.* University of California Press, Berkeley.

Stent, G. S. 1972. Prematurity and uniqueness in scientific discovery. *Scientific American,* Dec.: 84–93.

Stolzenburg, W. 2008. *Where the Wild Things Were: Life, Death, and Ecological Wreckage in a Land of Vanishing Predators.* Bloomsbury Press, New York.

Sunderland, M. E. 2012. Collections-based research at Berkeley's Museum of Vertebrate Zoology. *Historical Studies in the Natural Sciences* 42:83–113.

———. 2013. Teaching natural history at the Museum of Vertebrate Zoology. *British Journal for the History of Science* 46(1):97–121.

Sunquist, M. 1982. Unusual death of an Indian python (*Python molurus*). *Hornbill* 1:9.

Tantillo, J. A. 2001. Sport hunting, *Eudaimonia,* and tragic wisdom. *Philosophy in the Contemporary World* 8:101–112.

Taylor, B. 1891. *At Home and Abroad: A Sketch-Book of Life, Scenery, and Men.* G. P. Putnam's Sons, New York.

Taylor, E. H. 1952. *A Review of the Frogs and Toads of Costa Rica.* University of Kansas Science Bulletin 35(1). University of Kansas, Lawrence.

Thomas, E. M. 1990. The old way. *New Yorker,* Oct. 15: 78–110.

Tierny, J. 1987. Stephen Jay Gould: the *Rolling Stone* interview. *Rolling Stone,* Jan. 15: 38.

Tinbergen, N. 1963. On aims and methods of ethology. *Zeitschrift für Tierpsychologie* 20:410–429.

Trail, P. W. 1987. Predation and anti-predator behavior of Guianan cock-of-the-rock leks. *Auk* 104:496–507.

Truett, J., and M. Phillips. 2009. Beyond historic baselines: restoring bolson tortoises to Pleistocene ranges. *Ecological Restoration* 27:144–151.

Turner, J. 1996. *The Abstract Wild.* University of Arizona Press, Tucson.

Twigger, R. 1999. *Big Snake: The Hunt for the World's Longest Python.* Victor Gollancz, London.

Van Denburgh, J. 1922. *The Reptiles of Western North America: An Account of the Species Known to Inhabit California and Oregon, Washington, Idaho, Utah, Nevada, Arizona, British Columbia, Sonora, and Lower California.* Occasional Papers of the California Academy of Sciences 10.

Van Denburgh, J., and J. R. Slevin. 1918. *The Garter Snakes of Western North America.* Proceedings of the California Academy of Sciences, ser. 4, 8(6).

Van Devender, T. R., and G. L. Bradley. 1994. Late Quaternary amphibians and reptiles from Maravillas Canyon Cave, Texas, with discussion of the biogeography and evolution of the Chihuahuan Desert herpetofauna. Pp. 23–54 in P. R. Brown and J. W. Wright (eds.), *Herpetology of North American Deserts: Proceedings of a Symposium.* Southwestern Herpetological Society, Van Nuys CA.

Vonk, F. J., J. F. Admiraal, K. Jackson, R. Reshef, M. A. G. de Bakker, K. Vanderschoot, I. van den Berge, M. van Atten, E. Burgerhout, A. Beck, P. J. Mirtschin, E. Kochva, F. Witte, B. G. Fry, A. Woods, and M. K. Richardson. 2008. Evolutionary origin and development of snake fangs. *Nature* 454:630–633.

Vonk, F. J., K. Jackson, R. Doley, F. Madaras, P. J. Mirtschin, and N. Vidal. 2011. Snake venom: from fieldwork to the clinic. *BioEssays* 33:269–279.

de Waal, F. 2013. *The Bonobo and the Atheist.* W. W. Norton, New York.

Wallace, A. R. 1867. Mimicry and other protective resemblances among animals. *West Minster and Foreign Quarterly Review*, n.s., 32:1–43.

———. 1869. *The Malay Archipelago, the Land of the Orang-Utan and the Bird of Paradise: A Narrative of Travel with Studies of Man and Nature*. Macmillan, London.

Watson, J. D. 1968. *The Double Helix: A Personal Account of the Discovery of the Structure of DNA*. Atheneum, New York.

Westoff, G., M. Boetig, H. Cleckmann, and B. A. Young. 2010. Target tracking during venom "spitting" by cobras. *Journal of Experimental Biology* 213:1797–1802.

Wharton, C. H. 1966. Reproduction and growth in the cottonmouth, *Agkistrodon piscivorus* Lacepede, of Cedar Keys, Florida. *Copeia* 1966:149–161.

Wickler, W. 1968. *Mimicry in Plants and Animals*. McGraw-Hill, New York.

Williams, E. E., and W. E. Duellman. 1984. *Anolis fitchi*, a new species of the *Anolis aequatorialis* group from Ecuador and Colombia. Pp. 257–266 in R. A. Seigel et al. (eds.), *Vertebrate Ecology and Systematics: A Tribute to Henry S. Fitch*. University of Kansas Museum of Natural History Special Publication 10. University of Kansas, Lawrence.

Williams, T. 2004. Going catatonic. *Audubon*, Oct.

Wilson, E. O. 1975. *Sociobiology*. Harvard University Press, Cambridge MA.

———. 1999. Prologue. Pp. xi–xii in G. Daws and M. Fujita, *Archipelago: The Islands of Indonesia*. University of California Press, Berkeley.

Wright, A. H., and A. A. Wright. 1949. *Handbook of Frogs and Toads of the United States and Canada*. Comstock, Ithaca NY.

———. 1957. *Handbook of Snakes of the United States and Canada*. Comstock, Ithaca NY.

Wynn, E. J. 1944. *Bombers Across*. E. P. Dutton, New York.

Young, B. A., M. Boetig, and G. Westhoff. 2011. Functional bases of the spatial dispersal of venom during cobra "spitting." *Physiological and Biochemical Zoology* 82:80–89.

Zamma, K. 2011. Responses of chimpanzees to a python. *Pan Africa News* 18(2). Available at http://mahale.main.jp/PAN/18_2/18(2)_01.html (accessed Jan. 26, 2013).

INDEX

Species are listed by their common name (followed by the scientific name in parentheses) under arthropods; birds; crocodilians; fish; frogs; lizards; mammals; plants; salamanders; snakes; and turtles. Genera mentioned only by their scientific name in the text are indexed as such under these same groupings. Italic page references refer to illustrations.

birds *(continued)*
    ivory-billed woodpecker (*Campephilus
    principalis*), 237; keel-billed toucan (*Rham-
    phastos sulfuratus*), 113, 122; king vulture
    (*Sacroramphus papa*), 113; laughing falcon
    (*Herpetotheres cachinnans*), 181; long-eared
    owl (*Asio otus*), 52; mourning dove (*Zenaida
    macroura*), 44, 217–218; peregrine falcon
    (*Falco peregrinus*), 194; pintail (*Anas acuta*),
    211; quetzal (*Pharmomachrus mocinno*), 107;
    red-tailed hawk (*Buteo jamaicensis*), 43, 90,
    152; ringed kingfisher (*Megaceryle torquata*),
    113; squirrel cuckoo (*Piaya cayana*), 113;
    turkey vulture (*Cathartes aura*), 199, 214,
    218, 221, 223; and *Tyrannosaurus,* 194;
    vermillion flycatcher (*Pyrocephalus rubinus*),
    218, 240; wattled jacana (*Jacana jacana*), 136;
    white hawk (*Pseudastur albicollis*), 107; wild
    turkey (*Meleagris gallopavo*), 215, 251n4;
    yellow-billed magpie (*Pica nuttalli*), 13
Blair, W. Frank, 51, 68, 242n12
Blanchard, Frank N., 19
Bock, Carl E., 191–192
Bock, Jane H., 191–192, 194
Bowden, Charles, 236
Boyd, Carolyn E., 233–236
Brilliant Mary (longhorn cow), 206, 209
Brosnan, Sarah F., 189
Brown, William S. ("Bill"), 225
Browne, Jackson, 175
Buffalo Springfield (band), 177
Buffalo Springs (longhorn cow), 218
Burghardt, Gordon W., 75–83, 136, 187–189, 217,
    244n21
Burney, David A., 191–192
Butynski, Thomas M., 115, 119
Byrds (band), 177

Cadle, John E., 166
Caldwell, Janalee P., 175
California Academy of Sciences, 115, 149
Callicott, J. Baird, 202
Camp, Charles L., 16, 18
Campbell, Jonathan A., 73, 79, 151
Campbell, Joseph, 179
Camus, Albert, 84, 85, 231, 239, 245n33
Carothers, John H., 96
Carpenter, Charles C., 30–31
Carroll, C. Ronald, 83
Cézanne, Paul, 214
Cinco de Mayo (longhorn bull), 208–209, *210*,
    214, 240
Clapton, Eric, 177

Clark, Donald R., 47, 52, 108, 225
Clark, Rulon W., 225
Clutton-Brock, Juliet, 217
Cole, Charles J., 33–34, 60, 73–74
Colwell, Robert K., 91, 97, 104, 108, 150
Conant, Isabelle Hunt, 61
Conant, Roger, 12, 30, 32, 60–61, 170, 185
Cornell University Museum of Vertebrates, 139
Crazy Horse (Oglala Lakota), 17
Crick, Francis H. C., 13, 46, 233
crocodilians: American alligator (*Alligator
    mississippiensis*), 68, 185–186; American
    crocodile (*Crocodylus acutus*), 82, 185–186;
    Crocodylus anthrophagus (extinct man-
    eating crocodile), 186; Nile crocodile
    (*Crocodylus niloticus*), 186; spectacled
    caiman (*Caiman crocodilus*), 113, 136
Cundall, David, 139, 162–163

Darwin, Charles, 9, 11–13, 18, 22, 25, 69, 73, 79,
    90, 95, 180–181, 183, 186, 201, 229, 244n5
Davis, William B., 172
De Kay, James E., 181
Dial, Benjamin E., 32–33, 54, 169–179, 222, 233,
    236, 238
dinosaurs, 82, 113, 127, 186; *Tyrannosaurus,* 194
Dioum, Baba, 220, 223, 227, 251n1
Ditmars, Raymond L., 15–16, 27, 107
Dixon, James R., 172
Dobbs, Newt, 231
Dobie, J. Frank, 206
Dobrot, Steve, 192
Donlan, C. Josh, 191, 193
Dooley, Tom, 30
Drewes, Robert C., 115, 117, 119–120
Drummond, Hugh M., 83
Duellman, William E., 46
Dugan, Beverly A., 82–84
Dunn, Emmett Reid, 80
Duvall, Robert, 230
Dylan, Bob, 36, 84

Eagles (band), 175
Eakin, Richard M., 90
Echelle, Alice Fitch, 34, 44, 48–49, 52, 227–
    228, 230, 242n6
Echelle, Anthony A., 34, 49, 52, 230, 242n6
Echelle, Lena, 230
Echelle, Tyson, 230
Echternacht, Arthur C. ("Sandy"), 81, 89
Edwards, Cody W., 203–204, 206, 208
Ehrlich, Paul R., 46
Eibl-Eibesfeldt, Irenäus, 70

story palm), 109; *Iriartea* (stilt-palm), 109;
jicaro (aka monkey pot tree, *Lecythis ampla*),
110, 191; Joshua tree (*Yucca brevifolia*), 85–86,
94, 103, 176; lavender skeleton-plant (*Lygodes-
mia texana*), 239; loblolly pine (*Pinus taeda*),
68; mesquite (*Prosopis glandulosa*), 215, 217;
monkey pot tree (aka jicaro, *Lecythis ampla*),
110, 191; Osage orange (*Maclura pomifera*),
192; peyote cactus (*Lophophora williamsii*),
174; pickleweed (*Salicornia virginica*), 179;
saguaro (*Carnegiea gigantea*), 103; sycamore
(*Platanus occidentalis*), 164; white-thorn
acacia (*Acacia constricta*), 164; willow oak
(*Quercus phellos*), 68; yellow rock coreopsis
(*Coreopsis auriculata*), 239
Plummer, Michael H., 47
Pooch (cowboy), 238–239
Popper, Karl, 80
Prado (museum), 60
Pyburn, Karen Anne, 72, 244n1
Pyburn, Wanda C., 68, 72, 88, 244n1
Pyburn, William F. ("Bill"), 68–70, 72–75, 79,
80, 88, 94, 233, 244n1

Quammen, David, 177
Quannah (longhorn bull), 218

Radinsky, Leonard B., 81, 104–105
Rand, A. Stanley, 82, 114
Rawlins, Erle, 174–175, 177
Red Cloud (Oglala Lakota), 17
Redford, Robert, 98
Reinert, Howard K., 162–163, 225
Riechert, Susan E., 79–80
Riley (yellow lab), 11, 64, 186
Rivas, Jesús A., 136–137, 140, 188
Robbins, Tom, 128
Roemer, Gary W., 192
Rohlf, F. James, 46
Rossman, Douglas A., 24
Rumi (Persian poet), 85
Russell, Bertrand, 69
Russell, Ward, 20
Ruthven, Alexander G., 22–23, 242n13

salamanders: black (*Aneides flavipunctatus*), 92;
European fire (*Salamandra salamandra*),
60; grotto (*Eurycea spelea = Typhlotriton
speleaus*), 31, 242n4; Pacific giant (*Dicamp-
todon ensatus*), 92; red-bellied newt (*Taricha
rivularis*), 91; rough-skinned newt (*Taricha
granulosa*), 92; tiger (*Ambystoma tigrinum*),
31, 151; torrent (*Rhyacotriton variegatus*), 92

San Joaquin Experimental Range, 41–44,
51–52, 229
Santana, Manuel A., 163
Savage, Jay M., 48, 50
Savitzky, Alan H., 189
Schaller, George B., 6, 75, 183, 207
Schmidt, Karl P., 24, 71
Schneider, Christopher J., 203
Schnurrenberger, Hans, 149
Seigel, Richard A., 51
Senckenberg Museum of Natural History
(Naturmuseum Senckenberg), 60, 149
Shine, Richard, 53, 134–135, 244n10
Shonuff (longhorn cow), *210*
Shull, Elizabeth A., 83
Sigala-Rodríguez, J. Jesús, 11
Simon, Paul, 121
Sisyphus, 196
Skutch, Alexander F., 181–182, 202
Slevin, Joseph R., 22
Slowinski, Joseph B., 149
Smith, Arthur C. Jr., 226
Smith, Felisa A., 192
Smith, Hobart M., 13
Smith, John, 224, 251n6
Smith, Philip W., 47
Smith-Blackwell, Polly, 226
Smithsonian Institution, 13, 88, 252n35
Smithsonian Tropical Research Institute, 82–83
snakes: Andean short-tailed (*Tachymenis
peruviana*), 166, 249n34; Arizona black
rattlesnake (*Crotalus cerberus*), 166; Asian
rock python (*Python molurus*), 133, 137;
Australian brown (*Pseudonaja textilis*), 159;
Baird's ratsnake (*Pantherophis bairdi*), 173;
Baja California lyresnake (*Trimorphodon
lyrophanes*), 94, 96; banded rock rattlesnake
(*Crotalus lepidus klauberi*), 178; bandy-
bandy (*Vermicella annulata*), 159; bird
(*Pseustes poecilonotus*), 82; black mamba
(*Dendroaspis polylepis*), 147, 152, 157, *158*, 159;
black-necked gartersnake (*Thamnophis
cyrtopsis*), 88; black-necked spitting cobra
(*Naja nigricollis*), 147; black ratsnake
(*Pantherophis spiloides*), 187; black-tailed
rattlesnake (*Crotalus molossus*), 6–9, *7*, *9*,
60, 130–132, 163–166, 187, 190, 197, 201, 225,
241; Blair's kingsnake (*Lampropeltis blairi*),
see under gray-banded kingsnake; blotched
watersnake (*Nerodia erythrogaster trans-
versa*), 239; boa constrictor (*Boa constrictor*),
49, 137–138, 143; boomslang (*Dispholidus
typus*), 149, 154, 160; broad-banded